Advances in Solar Photovoltaic Energy Systems

Edited by Almoataz Y. Abdelaziz, Mahmoud A. Mossa and Najib El Ouanjli

Published in London, United Kingdom

Advances in Solar Photovoltaic Energy Systems
http://dx.doi.org/10.5772/intechopen.111228
Edited by Almoataz Y. Abdelaziz, Mahmoud A. Mossa and Najib El Ouanjli

Contributors
Enas Raafat Maamoun Shouman, Mukhammad-Sultan Payzullakhanov, Rasul Akbarov, Lola Suvonova, Ali Reza Reisi, Sundaramurthy Devikala, Johnson Maryleedarani Abisharani, Hamdy M. Sultan, Mahmoud A. Mossa, Almoataz Y. Abdelaziz

First published in London, United Kingdom, 2024 by IntechOpen
IntechOpen is the global imprint of INTECHOPEN LIMITED, registered in England and Wales, registration number: 11086078, 5 Princes Gate Court, London, SW7 2QJ, United Kingdom

British Library Cataloguing-in-Publication Data
A catalogue record for this book is available from the British Library

Additional hard and PDF copies can be obtained from orders@intechopen.com

Advances in Solar Photovoltaic Energy Systems
Edited by Almoataz Y. Abdelaziz, Mahmoud A. Mossa and Najib El Ouanjli
p. cm.
Print ISBN 978-0-85466-071-1
Online ISBN 978-0-85466-070-4
eBook (PDF) ISBN 978-0-85466-072-8

Meet the editors

Almoataz Y. Abdelaziz received a BSc and MSc in Electrical Engineering from Ain Shams University, Cairo, Egypt, in 1985 and 1990, respectively. He obtained a Ph.D. in Electrical Engineering according to the channel system between Ain Shams University, Egypt, and Brunel University, UK in 1996. He has been a Professor of Electrical Power Engineering at Ain Shams University since 2007. He has authored or co-authored more than 525 refereed journal and conference papers, 45 book chapters, and 6 edited books. His research areas include the applications of artificial intelligence, evolutionary, and heuristic optimization techniques to power systems operation, planning, and control. Prof. Abdelaziz is the chair of the IEEE Education Society chapter in Egypt, editor of *Electric Power Components and Systems*, and an editorial advisory board member for many international journals.

Mahmoud A. Mossa received bachelor's and master's degrees in electrical engineering from the Faculty of Engineering, Minia University, Egypt, in 2008 and 2013, respectively, He obtained a Ph.D. in Electrical Engineering from the University of Padova, Italy, in 2018. He is currently an associate professor in the Faculty of Engineering, Minia University, Egypt. He was a postdoctoral fellow at the Department of Industrial Engineering, University of Padova, Italy in 2021–2022. His research interests include renewable energy systems, power management, control systems, electric machine drives, power electronics, optimization, and load frequency control. He is an associate editor for the *International Journal of Robotics and Control Systems.* He is also a guest editor for special issues of *Energies* and *Machines*.

Najib El Ouanjli is a professor in the Department of Applied Physics, Hassan First University, Morocco. He received a master's degree from the Sidi Mohammed Ben Abdellah University Faculty of Sciences Dhar El Mahraz, Morocco, in 2015 and a Ph.D. from Sidi Mohammed Ben Abdellah University, Higher School of Technology of Fez, Morocco, in 2021. He was a Professor of Physics at the Science Ministry of Education from 2013 to 2021. His research interests include renewable energy systems, electrical and electronics engineering, rotating electric machines, system modeling, control techniques, optimization techniques, fault diagnosis, wind turbines, and solar energy. Dr. Ouanjli has published more than sixty-eight scientific publications. He is also a reviewer for international journals and conferences.

Contents

Preface **IX**

Chapter 1 **1**
Modeling Based on Daily Data of PV Power Plants
by Ali Reza Reisi

Chapter 2 **19**
Solar Power Prediction with Artificial Intelligence
by Enas Raafat Maamoun Shouman

Chapter 3 **47**
Parameter Identification of Solar Cell Mathematical Models Using Metaheuristic Algorithms
by Hamdy M. Sultan, Mahmoud A. Mossa and Almoataz Y. Abdelaziz

Chapter 4 **85**
Use of Concentrated Solar Power Technology for a High Temperature Processes: Case Study of Uzbekistan
by Mukhammad-Sultan Payzullakhanov, Rasul Akbarov and Lola Suvonova

Chapter 5 **105**
Addition of Organic Compounds in Gelatin-Biopolymer Gel Electrolyte for Enhanced Dye-Sensitized Solar Cells
by Sundaramurthy Devikala and Johnson Maryleedarani Abisharani

Preface

Due to greenhouse gas emissions, the world's growing energy consumption has inevitably resulted in increased atmospheric pollution and global warming. The world has started to establish decarbonization and sustainability targets that call for greater use of renewable energy sources since environmental protection has become a very critical issue. Solar energy conversion systems (SECS), like photovoltaic (PV) systems, are particularly interesting among these renewable energy sources since they are cost-effective and environmentally benign sources of electricity. In recent years, optimizing and lowering the cost of energy produced by PV systems has been a particularly popular issue. Efforts to reduce costs often take two forms: enhancing the materials and physical makeup of PV cells and utilizing power electronic circuits with the PV generator to increase the solar system's efficiency. Furthermore, random climatic factors, like temperature and irradiance flux, have a significant impact on PV system performances. As a result, modeling PV panels and creating optimization plans to maximize power extracted and boost efficiency under various irradiance circumstances are crucial tasks. This book provides a thorough review of new ideas and developments for solar PV energy systems by employing cutting-edge techniques.

The book is intended for scientists and engineers who are interested in the most recent developments in solar energy systems or the renewable energy sector. Professionals in the field of solar energy, such as research scientists, project managers, system operators, planners, engineers, investors, and financiers of solar energy projects, are specifically addressed.

The book includes five chapters. Chapter 1, "Modeling Based on Daily Data of PV Power Plants", investigates the modeling performance characteristics that are essential for the design and optimal operation of solar power plants. The target of this investigation is to determine an exact relationship between output power and weather conditions to avoid the influence of various factors on the performance of solar panels and data changes over time. A hybrid method based on genetic programming is presented for accurate modeling of solar power plan characteristics. The study evaluation is verified using a 3-kW solar power plant, and based on the obtained results, the effectiveness of the proposed modeling method is approved.

Chapter 2, "Solar Power Prediction with Artificial Intelligence", discusses optimizing renewable energy integration and ensuring efficient grid management through the application of artificial intelligence (AI) techniques for accurate solar power forecasting. The AI models considered include Artificial Neural Networks (ANNs), Support Vector Machines (SVMs), Random Forest (RF), and Gradient Boosting (GB). These models are selected based on their ability to capture complex patterns and non-linear relationships present in the solar energy data. Feature selection methods are utilized to identify the most relevant features that influence solar power generation. A comprehensive report detailing the forecasting process, methodology, and results is generated, allowing decision-makers to make informed choices based on the forecasted solar energy data.

Chapter 3, "Parameters Identification of Solar Cells Mathematical Models Using Metaheuristic Algorithms", focuses on simulating and assessing the performance of PV systems by constructing a precise mathematical representation using experimentally gathered data from solar cells and PV modules. Different metaheuristic algorithms are employed to deduce the unknown parameters inherent in various modes of solar cells and PV modules. Comprehensive statistical analysis is carried out to validate the efficacy and stability of the selected algorithms.

Chapter 4, "Use of Concentrated Solar Power Technology for a High Temperature Processes: Case Study of Uzbekistan", discusses the state of and prospects for the development of renewable energy use in Uzbekistan. It presents a comparative analysis of the technical and optical-energy characteristics of high-temperature solar furnaces of well-known research centers in Uzbekistan and their application in solving urgent scientific and technical problems. It illustrates some features of high-temperature processes by analyzing the process of quenching the considered high-temperature materials.

Chapter 5, "Addition of Organic Compounds in Gelatin-Biopolymer Gel Electrolyte for Enhanced Dye-Sensitized Solar Cells", investigates the performance enhancement of dye-sensitized solar cells via doping of a series of organic compound additives. It analyzes the effect of utilizing gelatin (GLN) biopolymer-based gel electrolytes on the ionic conductivity, stability, and ability of ionic mobility. The authors carry out a series of simulation and experimental tests to outline the role played by the added organic compounds.

We are grateful to everyone who helped with this book, including the editorial team, reviewers, and authors.

Almoataz Y. Abdelaziz
Faculty of Engineering,
Ain Shams University,
Cairo, Egypt

Mahmoud A. Mossa
Electrical Engineering Department,
Faculty of Engineering,
Minia University,
Minia, Egypt

Najib El Ouanjli
Faculty of Sciences and Technology,
Hassan First University,
Settat, Morocco

Chapter 1

Modeling Based on Daily Data of PV Power Plants

Ali Reza Reisi

Abstract

Modeling performance characteristics is essential for the design and optimal operation of solar power plants. However, due to the influence of various factors on the performance of solar panels and data changes over time, determining an exact relationship between output power and weather conditions is still challenging. In this chapter, a hybrid method based on genetic programming will be presented for accurate modeling of solar power plant characteristics, which includes two steps. First, three points of open-circuit voltage, maximum power point, and short-circuit current are modeled as functions of atmospheric conditions. For this purpose, by using the modeling process based on genetic programming, relationships with high fit will be obtained for these three points in terms of cell temperature and radiation. Then, with the help of these equations, the voltage–current characteristics are modeled based on the circuit analysis methods and without the need for factory data. To evaluate the modeling for a 3 kW solar power plant, and based on the results, the effectiveness of the proposed method will be shown.

Keywords: modeling, genetic programming, equivalent circuit, PV power plants, daily data

1. Introduction

Solar power plants based on photovoltaic systems are considered as one of the solutions to the energy crisis [1]. These power plants have been widely integrated into power systems due to their easy installation and utilization, low cost, and high lifetime [2]. This extensive integration, on a large scale, has brought challenges in consumption management and load-frequency control of microgrids because the production power of these power plants is intermittent and unplannable [3, 4]. In general, an accurate and reliable prediction of output power is of vital importance for the design and optimal operation of these power plants, which will lead to the improvement of grid stability and energy consumption efficiency.

Also, it is necessary to have an accurate model of the functional characteristics of solar power plants to detect faults, plan the time of washing the surface of the panels, identify loose electrical connections, and evaluate the performance of electronic power converters [5, 6]. For example, in fault diagnosis, after determining the voltage–current characteristic curve, if the working point of the power plant is at the point of maximum power, then the fault has not occurred and the control and power electronics, connections, and panels work correctly, but if the point If the power plant is not working at maximum power, an error has occurred, in this case, if the operating point is on the voltage–current characteristic, the error is related to sensors, electrical connections, or voltage converters, but if the power plant's operating point is on the voltage–current characteristic If there is no current, the fault is related to the solar panels.

So far, different methods have been presented for modeling solar panels [7]. Many PV power forecasting models have been developed in previous studies based on different strategies, such as data-driven modeling methods and equivalent circuit modeling. Different methods can be divided into three categories: 1) methods based on artificial intelligence, 2) methods based on equivalent circuits, and 3) combined methods.

In methods based on artificial intelligence, neural networks are usually used to model or identify the solar panel. Neural networks, although in topics such as fault detection [8], production energy prediction [9], estimation of some parameters such as solar cell temperature [10], amount of radiation on cloudy days [11], and losses due to dirt [12] perform well, the error of modeling the characteristics of solar panels is significant compared to other methods [13, 14]. In some research [14, 15], to improve accuracy, in addition to atmospheric conditions, voltage is also included as input and the output is only current or power. However, there was not much improvement in modeling accuracy. So, the modeling of solar panels based on these methods received less attention.

In methods based on the equivalent circuit, the solar panel is modeled by a non-linear circuit [16]. The main goal of these methods is to extract the unknown parameters of this non-linear circuit, which are dependent on atmospheric conditions. The algorithm of these methods is based on the methods of circuit analysis and the work data of the manufacturer [17, 18]. Complex circuit calculations are one of the challenges of these methods. In some of the proposed algorithms [19] some unknown parameters are assumed to be constant to reduce the calculation burden. Nonetheless, in practice, these parameters are variable, leading to a decrease in modeling accuracy [20, 21]. Another challenge of these methods is the dependence of calculations on the manufacturer's data (datasheet) [22]. These data change over time and therefore the accuracy of modeling in this method decreases. These cases are more frequent when dealing with the accuracy of the modeling of the solar power plant, which consists of interconnected solar panels.

The third category is the combined methods that do not suffer from the disadvantages of the previously mentioned methods. In combined methods, both equivalent circuits and neural networks are used to model the voltage–current characteristics of solar panels. In these methods [21, 23, 24], first, the neural network determines the unknown parameters according to atmospheric conditions. Then, the parameters are placed in the main equation of the non-linear equivalent circuit and the voltage–current characteristic can be provided. The main challenge of these methods is to form a database for training neural networks. A neural network needs a set of corresponding input and output data for training, but direct measurement of unknown parameters is practically not possible.

The importance of modeling accuracy is evident in the utilization process of a solar power plant that faces various challenges every day. Different methods presented for modeling the solar power plant are not free from challenges. In this regard, using the advantages of other modeling methods can be a solution to this problem.

Genetic programming (GP) is known as one of the most widely used tools in the field of modeling. GP can be considered a completed version of the genetic optimization algorithm as it is capable of extracting diverse and complex equations from databases and presenting them in the form of intuitive formulae [25]. Mohaghin used GP to solve various problems, including transportation energy demand prediction [26], analysis and optimization of thermocline solar energy storage [27], design and development of circuits and antennas [28], and identification of nonlinear systems [29]. Yet, so far, this powerful tool has not been adopted for modeling solar panels.

In this regard, in this chapter, a new method based on genetic programming for solar power plant modeling is presented, which is both accurate and independent of catalog data, so that it is possible to model old-installed solar power plants. The proposed method consists of two parts. In the first part, three main points of the voltage–current characteristic curve, i.e., open-circuit voltage, short-circuit current, and maximum power, are modeled according to weather conditions using the GP. In the second part, which is based on the equivalent circuit of a single diode, five unknown parameters of the equivalent circuit are calculated using the results of the first part and with the help of circuit analysis equations. In summary, the main contributions of this chapter are:

- A new hybrid method for modeling solar power plants based on daily data.
- Using genetic programming to model the solar power plant.
- Independent solar power plant modeling from datasheet.

The next sections of this chapter are organized as follows: In Section 2, the statement of the problem is discussed. The proposed method including two parts of GP and circuit analysis will be presented in Section 3. In Section 4, by simulating the proposed algorithm for a solar power plant, its performance is evaluated and the effective factors in the simulation are discussed. Finally, in Section 5, the general conclusions of the present study are summarized.

2. Statement of the problem

The solar cell, which consists of various materials such as silicon semiconductors, produces electricity from sunlight. A solar panel consists of several solar cells connected in series and parallel. **Figure 1** shows the equivalent circuit of a solar panel.

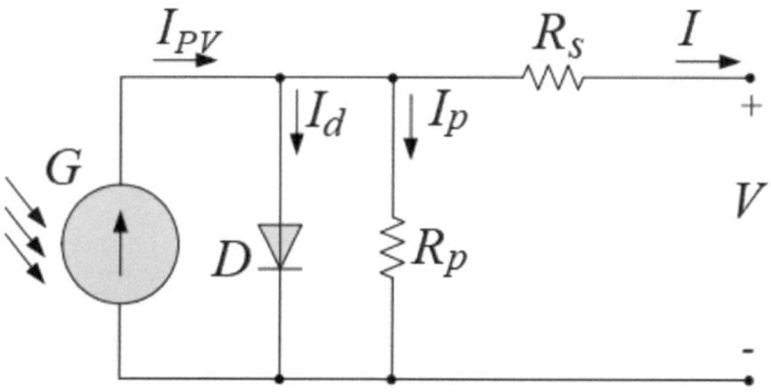

Figure 1.
Equivalent circuit of a single-diode solar panel.

Considering the parameters of **Figure 1** and circuit analysis, the characteristics of the solar panel are given here [30]:

$$I = I_{PV} - I_O\left[\exp\left(\frac{V + R_S I}{aV_t}\right) - 1\right] - \frac{V + R_S I}{R_P} \quad (1)$$

I_{PV} and I_O are related to the irradiance and temperature changes as follows [30]:

$$I_{PV} = (I_{S,n} + K_I \Delta T)\frac{G}{G_n} \quad (2)$$

$$I_O = \frac{I_{SC,n} + K_I \Delta T}{\exp(V_{OC,n} + K_V \Delta T)/aV_t - 1} \quad (3)$$

Additionally, Open-circuit voltage and short-circuit current are important points on the I-V characteristic curve of a solar panel. These points vary with changes in weather conditions. Using Eqs. (4) and (5), which are derived from the model equations, it is possible to calculate the short-circuit current and open-circuit voltage under different weather conditions [30].

$$I_{SC} = (I_{SC,n} + K_I \Delta T)\frac{G}{G_n} \quad (4)$$

$$V_{OC} = V_{OC,n} + K_V \Delta T \quad (5)$$

In the manufacturer's datasheet of a PV module, information such as open-circuit voltage (V_{OC}), short-circuit current (I_{SC}), maximum power value (Pm), current and voltage values corresponding to maximum power, Im and Vm, and temperature coefficients of current and voltage (K_I and K_V) are listed. However, the catalogs do not provide information about solar cell performance such as optical current (I_{PV}), saturation current (I_O), diode ideality factor (a), series resistance (R_S), and shunt resistance (R_P). These unknown parameters are essential for modeling a PV module.

So far, various methods have been presented for extracting unknown parameters. In these methods, parameter extraction is based on catalog data, such as K_I, K_V, $I_{SC,n}$, and $V_{OC,n}$, but the data for solar panels change over time and the modeling error increases. The challenge of combined methods in modeling solar panels is collecting data for neural network training, while the values of unknown parameters cannot be measured and must be extracted or calculated. In other words, it is not possible to model an old-installed solar power plant using analytical methods and combined methods.

In the next section, the proposed method for modeling the solar power plant is presented. The algorithm of the proposed method is independent of the factory datasheet, so old solar power plants can be properly modeled.

3. The proposed method

The suggested method for modeling the solar power plant includes two steps: 1) applying the GP, and 2) using circuit analysis. In the first step, three important points of voltage–current characteristics are modeled as formulae using the GP and according to cell temperature conditions and solar panel surface radiation. In the second step, the unknown parameters of the equivalent circuit are calculated based on the circuit analysis equations and using the formulae of the previous step. Each of the steps is described below.

3.1 Genetic programming

GP is a variant of genetic algorithm (GA) and was proposed (Koza [25]) to automatically code computer programs to perform predefined tasks. The GP method [25] is founded on the "survival of the fittest" and genetic propagation of characteristics principles followed by biologically evolving species. Although GA and GP employ the same principles of Darwinian evolution, there is a significant difference between their application domains. That is, while the GA performs function maximization/minimization, GP implements symbolic regression (SR). Given an example input–output data set, SR obtains an appropriate linear or a nonlinear function and all of its parameters that best fit the data.

The general form of the model obtained by GP-based SR for modeling is given as follows.

$$y = f(\alpha, \beta) \tag{6}$$

where f is a linear/non-linear function and α and β are the input values of the function. In this research, these values are the cell temperature and irradiance, the value of y, the output of the function, and one of the three values of open-circuit voltage, short-circuit current, and maximum power.

The stages of GP include four parts, which are performed by GeneXprotools.5 software in this research. The first part is the initialization, where the problem and the general structure of the solution are explained. The other three parts, namely, evaluation and selection of fit, intersection, and mutation are done by the software.

A summary of the GP procedure consisting of four major steps is given below.

i. Initialization: Creates a random initial population of candidate (probable) solutions to the given data-fitting problem using tree structures.

ii. Fitness evaluation and selection: Evaluate the fitness of each candidate solution in the current population using a fitness function and selects fitter solutions to form a pool of parent candidates (see **Figure 1b**) to undergo crossover.

iii. Crossover: Forms a new generation of candidate solutions (offspring); a typical crossover operation executed on a randomly selected pair from the parent pool is shown in **Figure 2a**.

iv. Mutation: Applies small changes to offspring candidate solutions (see **Figure 2b**).

Among these, steps (ii)–(iv) are performed iteratively until the best data-fitting candidate is achieved. Finally, statistical indicators are used to evaluate predictive models and compare modeling accuracy.

3.2 Circuit analysis

To plot the V-I characteristics of the solar panel, it is necessary to calculate or derive five parameters: a, I_O, I_{PV}, R_S, and R_P. Therefore, in this section, these

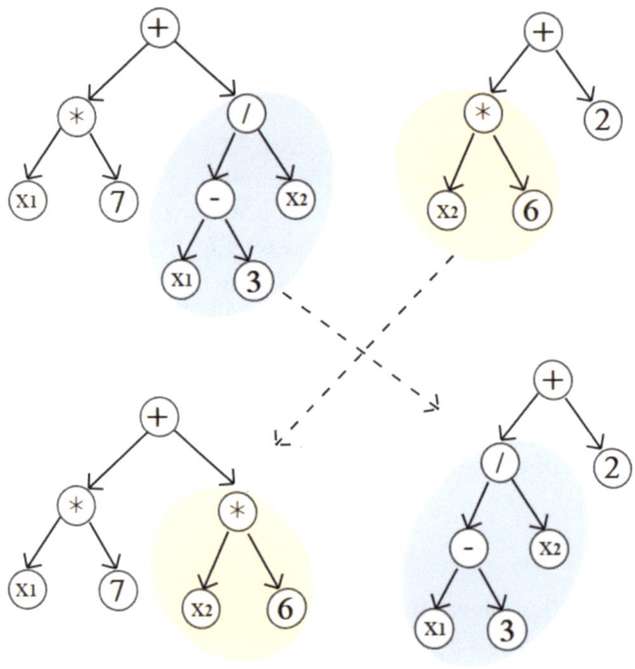

Figure 2.
Schematic of GP: (a) random selection of branches for reproduction, (b) crossover operation.

parameters are calculated using the results of the previous step and based on circuit analysis equations.

The coefficient a has a value between 1 and 2. Initially, its value is close to 1, and over time, and with the decrease in efficiency of the installed panels, its value approaches 2 [31]. Accordingly, the value of this coefficient is taken into account in the form of Eq. (16), which is the inverse of the fill factor,

$$a = \frac{I_{SC}^{GP} \times V_{OC}^{GP}}{P_m^{GP}} \tag{7}$$

By comparing the values of I_{sc}^{GP} and V_{oc}^{GP} with Eqs. (4) and (5) and substituting them in Eq. (3), we have,

$$I_O = \frac{I_{SC}^{GP}}{\exp\left(V_{OC}^{GP}\right)/aV_t - 1} \tag{8}$$

Considering that voltage and current values at the maximum power point are approximately seven-tenths of the open-circuit voltage and nine-tenths of the short-circuit current, $Vm \approx 0.7 \times V_{OC}$ and $Im \approx 0.9 \times I_{SC}$, [32] the initial values of resistances Rs and Rp can be found by the following equations,

$$R_{P0} = \frac{V_m}{I_{SC} - I_m} \approx 7 \times \frac{V_{OC}}{I_{SC}} \tag{9}$$

$$R_{SO} = \frac{V_{OC} - V_m}{I_m} \approx 0.34 \times \frac{V_{OC}}{I_{SC}} \tag{10}$$

Also, using the ratio $\frac{I_{PV,n}}{I_{SC,n}} = \frac{R_P + R_S}{R_P}$ [31], the following equation will replace Eq. (2),

$$I_{PV} = I_{SC}^{GP} \times \frac{R_P + R_S}{R_P} \tag{11}$$

To calculate the two R_S and R_P parameters, open-circuit voltage and short-circuit current values are inserted in Eq. (1),

$$0 = I_{PV} - I_O \left[\exp\left(\frac{V_{OC}}{aV_t} \right) - 1 \right] - \frac{V_{OC}}{R_P} \tag{12}$$

$$I_{SC} = I_{PV} - I_O \left[\exp\left(\frac{R_S I_{SC}}{aV_t} \right) - 1 \right] - \frac{R_S I_{SC}}{R_P} \tag{13}$$

Using Eqs. (7) to (13), the voltage–current characteristics curve is extracted based on the following algorithm.

1. Start
2. Calculate V_{OC}, I_{SC}, and Pm.
3. Calculate a using Eq. (7).
4. Calculate I_O using Eq. (8).
5. Set the initial values of R_P and R_S using Eqs. (9) and (10).
6. Calculate I_{PV} using Eq. (11).
7. Calculate R_P using Eq. (12).
8. Calculate Rs using Eq. (13).
9. Repeat steps 6–8 until the parameters converge.

Figure 3 shows the flowchart of the proposed method, which includes the above two steps to calculate the voltage–current characteristic curve.

4. Simulation and discussion

In this section, the performance of the proposed algorithm is evaluated for modeling a 3 kW power plant. For this purpose, first, by using GP and the data of a sunny day, open-circuit voltage, short-circuit current, and maximum power are determined in terms of solar cell temperature and radiation. Next, based on these formulae and using the circuit analysis, unknown parameters are extracted to model the voltage–current characteristic.

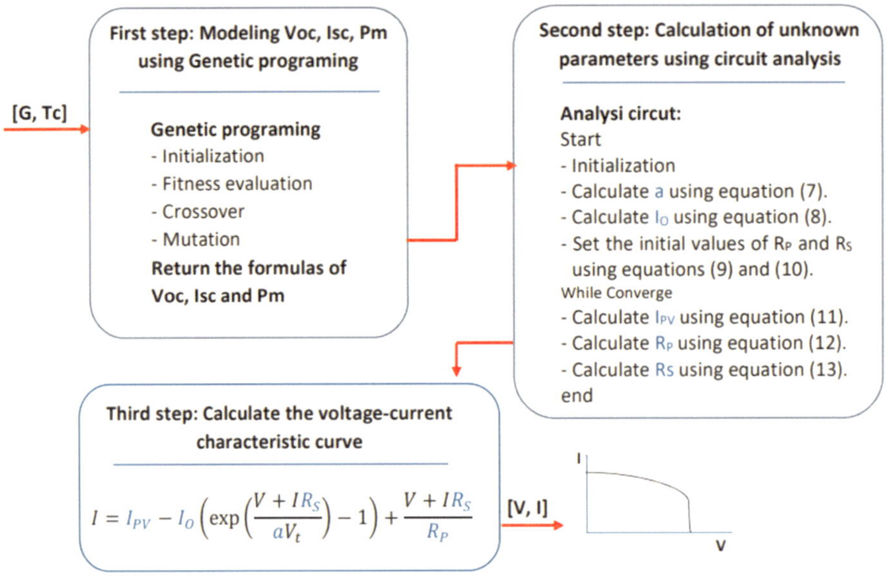

Figure 3.
The flowchart of the proposed method includes three steps: GP, circuit analysis, and V-I curve.

4.1 Genetic programming

In this section, short-circuit current, open-circuit voltage, and maximum power of a 3 kW power plant are modeled by GP according to cell temperature and solar panel surface radiation. The output of this section includes three formulae that can be used to calculate these three quantities for all cell temperatures and radiations. The initial structure of the formula of these three quantities is considered according to Eqs. (14)–(16).

$$I_{sc} = f_{pv_1}(T_C, G) \tag{14}$$

$$V_{oc} = f_{pv_2}(T_C, G) \tag{15}$$

$$P_m = f_{pv_3}(T_C, G) \tag{16}$$

After the implementation of GP, with the initial population size of 500, the number of generations of 1000, the crossover rate of 0.8, and the mutation probability of 0.12, the three suggested formulae for estimating the values of ISC, VOC, and Pm are:

$$I_{sc}^{GP} = -0.253 + 0.0047\, T_C + 0.0303\, T_C^{0.022} \times G \tag{17}$$

$$V_{oc}^{GP} = -222.07 - 0.57\, T_C + 308.52\, \mathrm{G}^{0.017} + 0.044\, T_C \times \mathrm{G}^{0.17} \tag{18}$$

$$P_m^{GP} = \left(-0.294 + 0.0294 T_C^{0.0047} \times G\right)\Big(66.43 - 0.237\mathrm{G}^{0.67} - 0.536\, T\, \mathrm{G}^{-0.044} + 6.144\, T_C^{0.011} \times \mathrm{G}^{0.325}\Big) \tag{19}$$

Two statistical indices of root mean square error (RMSE) and coefficient of determination, R2, according to the following equations are used to evaluate the accuracy of the proposed formulae.

$$RMSE = \sqrt{\frac{\sum_{i=1}^{N} \left(y_o - y_p \right)^2}{N_{sa}}} \tag{20}$$

$$R^2 = 1 - \frac{\sum_{i=1}^{N} \left(y_o - y_p \right)^2}{\sum_{i=1}^{N} \left(y_o - \bar{y} \right)^2} \tag{21}$$

Statistical indices including RMSE and R2 are reported in **Table 1**, Appendix A.

4.2 Circuit analysis

In this section, according to the proposed algorithm, **Figure 3**, having the formulae of open-circuit voltage, short-circuit current, and maximum power, the values of five unknown parameters can be calculated. **Table 2** shows the simulation results for the test data, Appendix B.

To check the performance of the proposed method, the modeling for other data should also be evaluated. For this purpose, the modeling of the solar power plant for a sunny day with cell temperature and radiation intensity shown in **Figure 4** has been done, **Figure 5** shows the performance of the proposed method.

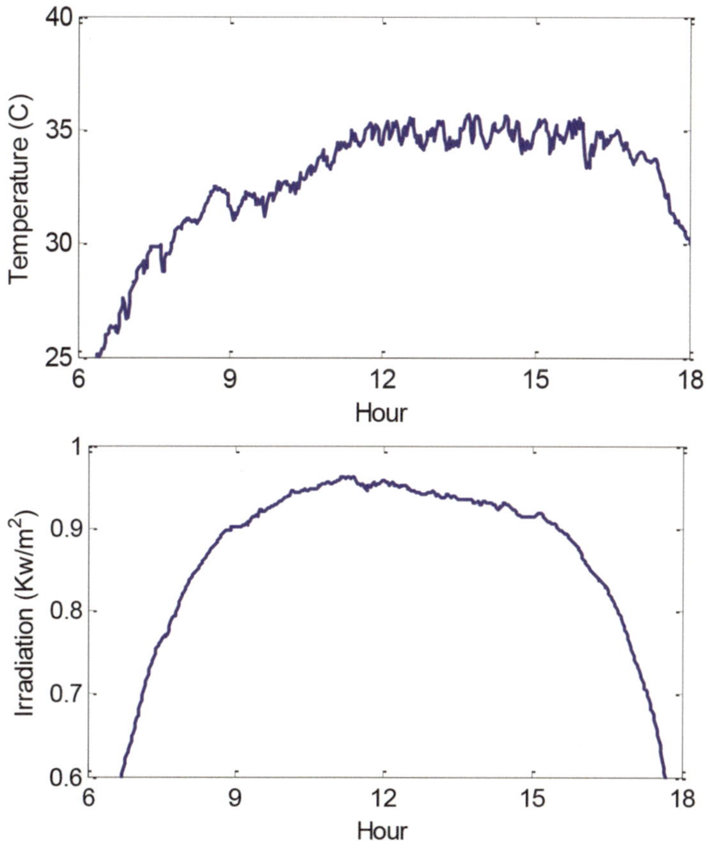

Figure 4.
Atmospheric conditions. a) Temperature, b) irradiance.

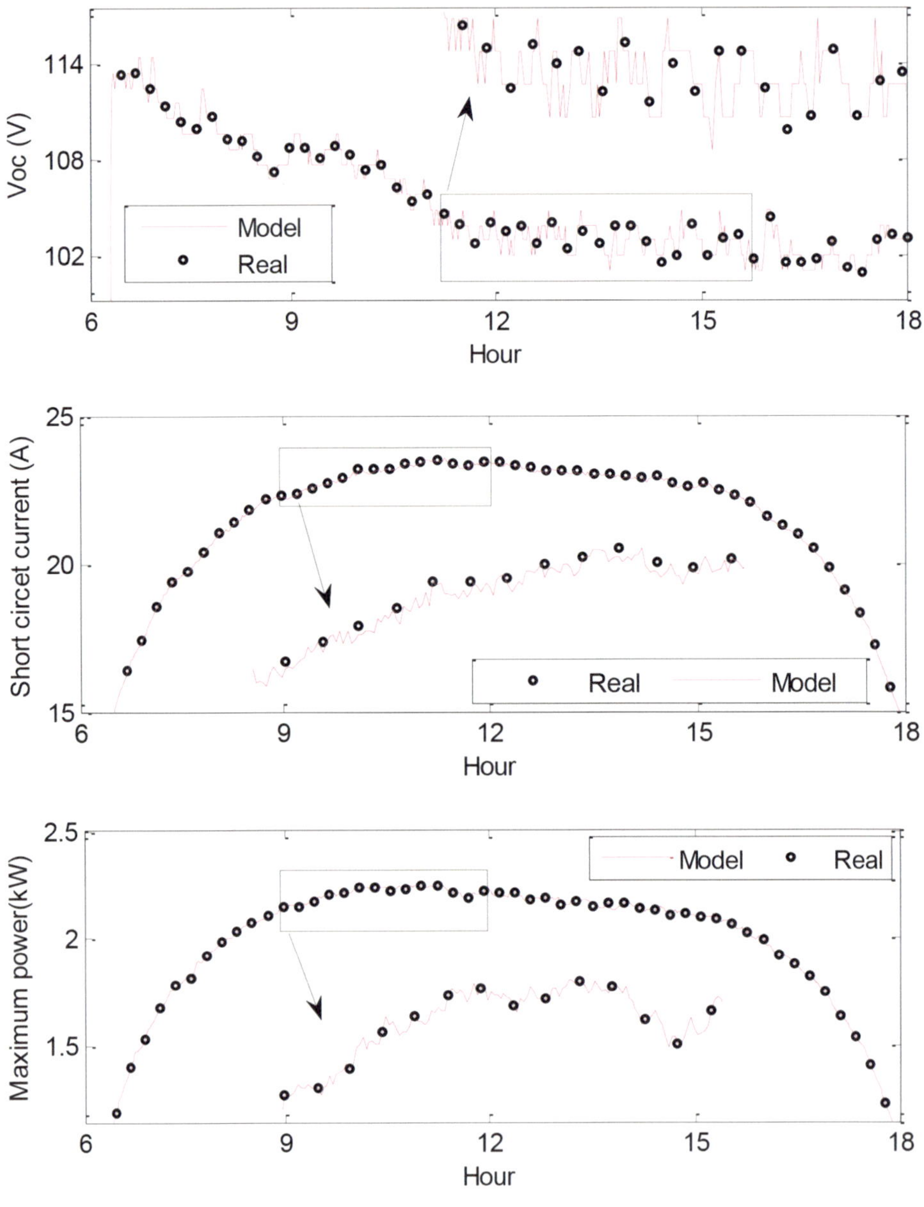

Figure 5.
Solar power plant modeling a) open-circuit voltage, b) short-circuit current, c) maximum power.

Figure 6 shows the voltage–current characteristics. Also, the values of RMSE and R2 are presented in **Table 3**, Appendix C. As can be seen, the results match the initialization data very accurately, which is due to the accurate modeling of five parameters based on the proposed equations. In this case, the parameters change depending on atmospheric conditions, such as the parameters a, Rp, and Rs, which are usually included in fixed circuit analysis methods. Here, based on the proposed formulae, the parameters change proportionally.

The proposed method can be compared with the hybrid method presented in the article [23]. This comparison can be made in the two fields of modeling process and

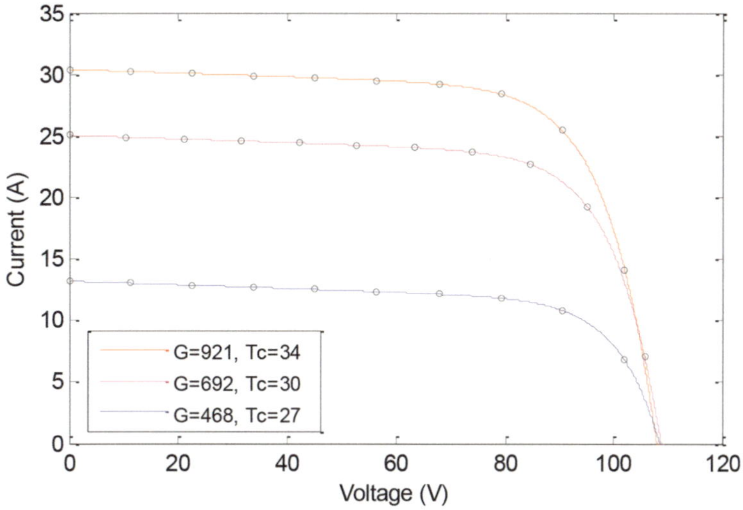

Figure 6.
Voltage–current characteristic of a 3 kW power plant for actual and modeled values.

modeling results. In the modeling process, databases play an effective role. In the proposed method, the databases are small and the required data can be measured, but the required data of the [23] method are five unknown parameters that cannot be measured directly, and these data are extracted with calculations and optimization methods. Also, to train the neural network in the method, the databases should be wider.

Another area where two methods can be compared is comparing their results. In this regard, the method of the [23] is implemented by the data of **Table 2**, so that part of the data is used for training and part as a test. The results of the simulations for the test data are shown in **Table 4**, Appendix D. The main reason for the low accuracy of the modeling done is the small amount of training data.

5. Discussion

In the utilization of the solar power plant, modeling the instantaneous values of three parameters: maximum power, open-circuit voltage, and short-circuit current are important. These values are effective both in the process of controlling and extracting power and in evaluating the performance of the controller. In the following, the effective factors in modeling and determining these three points are discussed in different ways.

5.1 The effect of time

The parameters and coefficients of solar panels change over time. These changes are the main challenge of the analytical methods because, in these methods, the parameters are extracted based on the factory data. The passage of time does not play a role in optimization methods because the parameter extraction algorithm is re-executed when the weather conditions change. In the modeling with the proposed algorithm method, the passage of time is taken into account in such a way that after each periodical maintenance service, once or twice a year, older data is deleted, newer data is added, and the formula is renewed. On other days of the year, modeling is performed based on the calculated formulae.

5.2 The effect of data type

In the analytical methods, the data required for modeling are diverse and include factory data (solar panel catalog), three important characteristic points of voltage–current, and atmospheric conditions. After modeling, the input data is the atmospheric condition and the output data is the voltage–current characteristic. With this feature, it is possible to extract three important points. In modeling by optimization method, the required data include different voltage–current points in certain atmospheric conditions. The output of this method is the five unknown parameters of Eq. (1). In other words, the voltage–current characteristic in specific atmospheric conditions it is possible to determine three important points. In the proposed method, the required data includes a set of three important points and weather conditions for two or three days of the year. After modeling the input data, atmospheric conditions and output data are three important points. Also, based on the proposed formulae, it is possible to extract five unknown parameters by analytical method without the need for a factory datasheet. Simply put, the proposed method is a combination of two analytical and optimization methods.

5.3 Fault analysis

Among these three methods, analytical methods provide less accuracy. In analytical methods, three out of five unknown parameters usually have a fixed value for different weather conditions. In other words, these three constant parameters are determined for STC; therefore, in other atmospheric conditions for low irradiance, modeling is associated with errors. In large power plants, including series and parallel panels, this error is more visible. In the optimization methods, the error has been reported as negligible because the optimization algorithm is implemented for each weather condition. Due to the continuous implementation of algorithms in practice, there is a possibility of measurement errors and hardware failure. In these methods, there is a compromise between the speed and accuracy of the response and the hardware used. The error of the proposed method depends on the quality of the collected data. If the data includes both sunny and cloudy days, the modeling error is negligible. Also, due to the functional changes of solar panels over time, older data should be excluded from the process of modeling and calculation of formulae.

6. Conclusions

Accurate modeling of the solar power plant is important both from the point of view of the power plant to diagnose faults, plan maintenance, and improve performance and from the smart networks to manage production and consumption and control frequency and stability. In this chapter, using the GP technique, three important points of voltage–current characteristics are modeled as formulae of atmospheric conditions. To calculate the proposed formulae, the data from one day at a solar power plant including atmospheric conditions are used as variables and three important points are adopted as outputs. It was also shown in this chapter that with the help of the calculated formulae, it is possible to extract five unknown parameters of the single-diode model with high accuracy and without the need for factory data. Finally, the proposed method was simulated for a 3 kW power plant. The simulation results were effective and the proposed method showed three important points of voltage–current characteristics while extracting the unknown parameters of the single-diode model.

Highlights

- A hybrid method for modeling the solar power plant will be investigated, which is simple and efficient.
- The proposed method is independent of the data sheet information, as it enables the modeling of panels that have been installed for several years.
- The method reviewed in this chapter performs modeling with high accuracy based on the daily data of the solar power plant, the required data are small and measurable.

Appendices and nomenclature

STC	the standard test condition (Tn = 25°C and Gn = 1000 W/m^2)
Tn	the solar cell temperature at standard test conditions (25°C)
T_C	the cell temperature (°C)
ΔT	the cell temperature difference from STC
Gn	the solar irradiance at standard test conditions (1000 W/m^2)
G	the solar irradiance (W/m^2)
$I_{PV,n}$	the photovoltaic current at standard test conditions (A)
I_{PV}	the photovoltaic current (A)
$V_{OC,n}$	the open circuit voltage at STC (V)
V_{OC}	the open circuit voltage (V)
$I_{SC,n}$	the short circuit current at STC (A)
I_{SC}	the short circuit current (A)
I_O	the reverse saturation current
V_t	the thermal voltage which is equal to $N_S K T_C / q$ (V)
N_S	the number of cells connected in a series
K	the Boltzmann constant
q	the electron charge
a	the diode ideality factor
R_S	the series resistances
R_P	the parallel resistances
K_I	the ratio of short-circuit current to temperature
K_V	the temperature coefficient of open-circuit voltage
N_{sa}	the number of samples
y_O	the data observed from practical experiments
y_P	the corresponding data obtained using the GP
$\bar{y}$	the mean of the data

Appendix A

See **Table 1**

Index	V_{OC}	I_{SC}	Pm
RMSE	0.0154	0.0162	0.0236
R2	0.9998	0.9997	0.9988

Table 1.
Statistical indices corresponding to each proposed equation.

Appendix B

See **Table 2**

GP input		GP output			Circuit analysis results				
T (W/m^2)	Tc (Co)	VOC (V)	ISC (A)	Pm (kW)	a	IPV (A)	IO (μA)	Rs (Ω)	Rp (Ω)
302.34	16.8	110.28	9.5739	0.7533	1.40	9.82	0.030	3.57	139.2
307.91	17.2	110.21	9.7605	0.7671	1.40	10.01	0.032	3.57	139.0
309.25	18.1	109.83	9.8192	0.7674	1.41	10.07	0.039	3.56	138.3
311.73	18.8	109.56	9.9109	0.7715	1.41	10.17	0.044	3.56	137.8
409.36	20.9	110.30	13.1071	1.0270	1.41	13.44	0.060	3.55	137.7
427.01	22.0	110.07	13.6997	1.0692	1.41	14.05	0.073	3.54	137.1
544.91	26.2	109.75	17.6112	1.3616	1.42	18.07	0.149	3.52	135.8
598.33	27.4	109.82	19.3753	1.4957	1.42	19.88	0.182	3.51	135.6
618.37	27.9	109.81	20.0387	1.5453	1.42	20.56	0.199	3.50	135.5
687.09	29.5	109.80	22.3141	1.7147	1.43	22.89	0.263	3.49	135.7
815.44	34.9	108.58	26.6278	2.0007	1.45	27.32	0.669	3.46	133.4
831.48	34.1	109.06	27.1356	2.0499	1.44	27.84	0.588	3.46	133.6
882.79	35.2	108.98	28.8412	2.1707	1.45	29.59	0.715	3.45	133.4
867.38	35.9	108.56	28.3518	2.1239	1.45	29.09	0.799	3.45	132.8
894.87	36.2	108.64	29.2598	2.1910	1.45	30.02	0.846	3.44	132.9
923.39	36.9	108.56	30.2110	2.2560	1.45	30.99	0.957	3.43	132.7
948.13	37.4	108.52	31.0342	2.3133	1.46	31.84	1.047	3.43	132.6
972.29	38.2	108.35	31.8455	2.3652	1.46	32.67	1.201	3.42	132.2

Table 2.
Simulation results for test data.

Appendix C

See **Table 3**

DOI: http://dx.doi.org/10.5772/intechopen.1003234

Index	V_{OC}	I_{SC}	Pm
RMSE	0.0213	0.0197	0.0251
R^2	0.9989	0.9991	0.9983

Table 3.
Statistical indices corresponding to each proposed equation.

Appendix D

See **Table 4**

Index	a	I_{PV} (A)	I_O (μA)	R_S (Ω)	R_P (Ω)
RMSE	0.1316	0.3548	0.3827	0.2161	0.1925
R^2	0.8485	0.6971	0.7283	0.7935	0.8218

Table 4.
Statistical indices corresponding to each proposed equation.

Author details

Ali Reza Reisi
Department of Electrical Engineering, Technical and Vocational University (TVU), Tehran, Iran

*Address all correspondence to: reisi.alireza@gmail.com

References

[1] Jordan DC, Kurtz SR. Field performance of 1.7 GW of photovoltaic systems. IEEE Journal of Photovoltaics. 2015;**5**(1):243-249. DOI: 10.1109/JPHOTOV.2014.2361667

[2] Brenna M, Foiadelli F, Longo M, Zaninelli D. Energy storage control for dispatching photovoltaic power. IEEE Transactions on Smart Grid. 2018;**9**(4): 2419-2428. DOI: 10.1109/TSG.2016.2611999

[3] Kamil K, Chong KH, Hashim H. Excess power rerouting in the grid system during high penetration solar photovoltaic. Electric Power Systems Research. 2023;**214**(Part A):108871. DOI: 10.1016/j.epsr.2022.108871

[4] Chao P, Li W, Liang X, Shuai Y, Sun F, Ge Y. A comprehensive review on dynamic equivalent modeling of large photovoltaic power plants. Solar Energy. 2020;**210**:87-100. DOI: 10.1016/j.solener.2020.06.051

[5] Hara S, Douzono H, Imamura M, Yoshioka T. Estimation of photovoltaic cell parameters using measurement data of photovoltaic module string currents and voltages. IEEE Journal of Photovoltaics. 2022;**12**(2):540-545. DOI: 10.1109/JPHOTOV.2021.3135262

[6] Ma M, Zhang Z, Yun P, Xie Z, Wang H, Ma W. Photovoltaic module current mismatch fault diagnosis based on I-V data. IEEE Journal of Photovoltaics. 2021;**11**(3):779-788. DOI: 10.1109/JPHOTOV.2021.3059425

[7] Garud K, Jayaraj S, Lee M-Y. A review on modeling of solar photovoltaic systems using artificial neural networks, fuzzy logic, genetic algorithm, and hybrid models. International Journal of Energy Research. 2021;**45**(1):6-35. DOI: 10.1002/er.5608

[8] Laurino M, Piliougine M, Spagnuolo G. Artificial neural network based photovoltaic module diagnosis by current–voltage curve classification. Solar Energy. 2022;**236**(1):383-392. DOI: 10.1016/j.solener.2022.02.039

[9] Huang C-J, Kuo P-H. Multiple-input deep convolutional neural network model for short-term photovoltaic power forecasting. IEEE Access. 2019;**7**:74822-74834. DOI: 10.1109/ACCESS.2019.2921238

[10] Khaleel FM, Hasan IA, Mohammed MJ. PV panel system modeling method based on the neural network. AIP Conference Proceedings. 2022;**2386**:040029. DOI: 10.1063/5.0066820

[11] Si Z, Yixiao Y, Yang M, Li P. Hybrid solar forecasting method using satellite visible images and modified convolutional neural networks. IEEE Trans. on Industry Applications. 2021;**57**(1):5-16. DOI: 10.1109/TIA.2020.3028558

[12] Zhang W, Liu S, et al. Deep-learning-based probabilistic estimation of solar PV soiling loss. IEEE Transactions on Sustainable Energy. 2021;**12**(4):2436-2444. DOI: 10.1109/TSTE.2021.3098677

[13] Almonacid F, Rus C, Hontoria L, Fuentes M, Nofuentes G. Characterisation of Si-crystalline PV modules by artificial neural networks. Renewable Energy. 2009;**34**(4):941-949. DOI: 10.1016/j.renene.2008.06.010

[14] Bonanno F, Capizzi G, Graditi G, Napoli C, Tina GM. A radial basis function neural network-based approach for the electrical characteristics estimation of a photovoltaic module.

Applied Energy. 2012;**97**:956-961. DOI: 10.1016/j.apenergy.2011.12.085

[15] Lopez-Guede JM, Ramos-Hernanz, Zulueta E, Fernandez-Gamiz U, Azkune G. Dual model oriented modeling of monocrystalline PV modules based on artificial neuronal networks. International Journal of Hydrogen Energy. 2017;**42**(28):18103-18120. DOI: 10.1016/j.ijhydene.2017.02.062

[16] Shongwe S, Hanif M. Comparative analysis of different single-diode PV modeling methods. IEEE Journal of Photovoltaics. 2015;**5**(3):938-946. DOI: 10.1109/JPHOTOV.2015.2395137

[17] Celik AN, Acikgoz N. Modelling and experimental verification of the operating current of mono-crystalline photovoltaic modules using four- and five-parameter models. Applied Energy. 2007;**84**(1):1-15. DOI: 10.1016/j.apenergy.2006.04.007

[18] Brano VL, Orioli A, Ciulla G, Di Gangi A. An improved five-parameter model for photovoltaic modules. Solar Energy Materials and Solar Cells. 2010;**94**(8):1358-1370. DOI: 10.1016/j.solmat.2010.04.003

[19] Ciulla G, Brano VL, Di Dio V, Cipriani G. A comparison of different one-diode models for the representation of I–V characteristic of a PV cell. Renewable and Sustainable Energy Reviews. 2014;**32**:684-696. DOI: 10.1016/j.rser.2014.01.027

[20] Hsieh Y-C, Li-Ren Y, Chang T-C, Liu W-C, Tsung-Hsi W, Moo C-S. Parameter identification of one-diode dynamic equivalent circuit model for photovoltaic panel. IEEE Journal of Photovoltaics. 2020;**10**(1):219-225. DOI: 10.1109/JPHOTOV.2019.2951920

[21] Dong X-J, Shen J-N, Ma Z-F, He Y-J. Simultaneous operating temperature and output power prediction method for photovoltaic modules. Energy. 2022;**260**:124909. DOI: 10.1016/j.energy.2022.124909

[22] Mahmoud Y, Xiao W. Evaluation of shunt model for simulating photovoltaic modules. IEEE Journal of Photovoltaics. 2018;**8**(6):1818-1823. DOI: 10.1109/JPHOTOV.2018.2869493

[23] Karatepe E, Boztepe M, Colak M. Neural network based solar cell model. Energy Conversion and Management. 2006;**47**:1159-1178. DOI: 10.1016/j.enconman.2005.07.007

[24] Celik AN. Artificial neural network modeling and experimental verification of the operating current of mono-crystalline photovoltaic modules. Solar Energy. 2011;**85**:2507-2517. DOI: 10.1016/j.solener.2011.07.009

[25] Koza JR. Genetic programming as a means for programming computers by natural selection. Statistics and Computing. 1994;**4**(2):87-112. DOI: 10.1007/BF00175355

[26] Mehdi Forouzanfar A, Doustmohammadi SH, Shakouri H. Transport energy demand forecast using multi-level genetic programming. Applied Energy. 2012;**91**(1):496-503. DOI: 10.1016/j.apenergy.2011.08.018

[27] Ghezelbash G, Babaelahi M, Saadatfar M. New analytical solution and optimization of a thermocline solar energy storage using differential quadrature method and genetic programming. Journal of Energy Storage. 2022;**52**(Part B):104806. DOI: 10.1016/j.est.2022.104806

[28] Clemens S, Iskander MF, Yun Z, Rayno J. Hybrid genetic programming

for the development of metamaterials designs with improved characteristics. IEEE Antennas and Wireless Propagation Letters. 2018;**17**(3):513-516. DOI: 10.1109/LAWP.2018.2800057

[29] dos Santos Coelho L, Bora TC, Klein CE. A genetic programming approach based on Lévy flight applied to nonlinear identification of a poppet valve. Applied Mathematical Modelling. 2014;**38**(5–6):1729-1736. DOI: 10.1016/j.apm.2013.09.014

[30] Moradi MH, Reisi AR. A hybrid maximum power point tracking method for photovoltaic systems. Solar Energy. 2011;**85**:2965-2976. DOI: 10.1016/j.solener.2011.08.036

[31] Humada AM, Darweesh SY, Mohammed KG, Kamil M, Mohammed SF, Kasim NK, et al. Modeling of PV system and parameter extraction based on experimental data: Review and investigation. Solar Energy. 2020;**199**:742-760. DOI: 10.1016/j.solener.2020.02.068

[32] Reisi AR, Moradi MH, Jamasb S. Classification and comparison of maximum power point tracking techniques for the photovoltaic system: A review. Renewable and Sustainable Energy Reviews. 2013;**19**:433-443. DOI: 10.1016/j.rser.2012.11.052

Chapter 2

Solar Power Prediction with Artificial Intelligence

Enas Raafat Maamoun Shouman

Abstract

Solar power prediction is a critical aspect of optimizing renewable energy integration and ensuring efficient grid management. The chapter explore the application of artificial intelligence (AI) techniques for accurate solar power forecasting. The AI models considered include Artificial Neural Networks (ANN), Support Vector Machines (SVM), Random Forest, and Gradient Boosting. These models are selected based on their ability to capture complex patterns and non-linear relationships present in the solar energy data. The solar power forecasting process involves data preprocessing, feature selection, model training, and evaluation. Data preprocessing techniques are applied to handle missing values and normalize the data to improve model performance. Feature selection methods are utilized to identify the most relevant features that influence solar power generation. The AI models are trained using historical data, where they learn the relationships between input features and solar power generation. Model evaluation is carried out using metrics such as Mean Squared Error (MSE) and Root Mean Squared Error (RMSE) to assess the accuracy of the forecasts. Furthermore, the forecasted results are visualized through line plots and error plots to provide valuable insights to stakeholders. A comprehensive report detailing the forecasting process, methodology, and results is generated, allowing decision-makers to make informed choices based on the forecasted solar energy data.

Keywords: solar power, artificial intelligence, data analysis, machine learning, statistical prediction methods

1. Introduction

While solar energy presents numerous benefits, its integration into the electricity grid introduces challenges related to its inherent intermittency and variability. Unlike conventional power sources like coal or natural gas plants that can produce a consistent output, solar power production fluctuates based on weather conditions, time of day, and geographical location. These fluctuations make it challenging for grid operators to maintain grid stability and ensure a reliable power supply.

Solar power forecasting plays a crucial role in addressing the challenges posed by solar energy's intermittent nature. Accurate solar power forecasts enable grid operators to anticipate fluctuations in solar generation and plan grid operations accordingly. By having precise forecasts of solar power production, utilities can optimize the use of solar energy while efficiently managing conventional power generation sources for backup.

Effective solar power forecasting contributes to grid stability and reduces the need for fossil fuel-based backup power generation. Integrating solar power forecasts into energy management systems allows for better coordination of electricity generation, distribution, and demand response strategies. It enables grid operators to optimize the dispatch of power resources, thus minimizing energy wastage and reducing overall operational costs.

The ability to predict solar power production aids in avoiding imbalances between forecasted and actual generation, which can lead to financial penalties or missed revenue opportunities.

Governments worldwide are increasingly setting renewable energy targets and implementing supportive policies to promote the adoption of solar power and other clean energy sources. Solar power forecasting is a critical component of these policies, as it helps grid operators meet renewable energy integration mandates and transition towards a low-carbon energy future.

In recent years, significant advancements in forecasting technology have improved the accuracy of solar power forecasts. Advanced weather modeling, artificial intelligence, machine learning algorithms, and high-resolution satellite imagery have enhanced the precision and lead times of solar power predictions.

In an era marked by the escalating demand for clean and sustainable energy sources, solar power has emerged as a promising solution to address the global energy challenge. As the world seeks to transition from fossil fuels to renewable resources, the efficient integration of solar energy into power systems assumes paramount importance. This necessitates the development of accurate solar power prediction models that enable optimal energy management and grid operation. The fusion of artificial intelligence (AI) techniques with solar power forecasting holds tremendous potential in realizing this objective.

This chapter delves into the realm of solar power prediction, focusing on the application of AI methodologies to enhance the accuracy and reliability of solar power forecasts. The underlying premise is that accurate predictions enable stakeholders to make informed decisions, thereby facilitating the seamless integration of solar energy into the broader energy landscape. By employing AI models, such as Artificial Neural Networks (ANN), Support Vector Machines (SVM), Random Forest, and Gradient Boosting, this chapter explores how intricate patterns and non-linear relationships inherent in solar energy data can be effectively captured.

Solar power forecasting, a multifaceted process, encompasses several stages aimed at transforming raw data into actionable insights. These stages include data preprocessing, feature selection, model training, evaluation, and visualization. Data preprocessing techniques are crucial to handle missing values and normalize data, thereby enhancing the performance of AI models. Feature selection methodologies contribute to identifying the most influential variables that impact solar power generation, streamlining the forecasting process.

The AI models employed in solar power prediction are trained using historical data, enabling them to discern intricate connections between input features and solar power output. The process involves extracting underlying trends and correlations, a task ideally suited to AI's prowess in pattern recognition. Model evaluation is paramount to ascertain the accuracy of predictions. Metrics such as Mean Squared Error (MSE) and Root Mean Squared Error (RMSE) serve as benchmarks to gauge the fidelity of forecasts, ensuring that the developed models meet rigorous performance standards.

Visualizing the forecasted results adds an essential layer of comprehension for stakeholders. Line plots and error plots offer intuitive insights, facilitating a clear understanding of the forecast's reliability and potential areas for improvement. The visualization component bridges the gap between technical complexity and practical applicability, empowering decision-makers to make informed choices based on the forecasted solar energy data.

Moreover, the culmination of these efforts is a comprehensive report, detailing the forecasting process, methodologies employed, and the resultant outcomes. This report empowers decision-makers and stakeholders with a comprehensive understanding of the forecast's underpinnings, enabling them to navigate the intricacies of solar power integration with confidence.

In summary, the marriage of AI techniques with solar power prediction presents an unprecedented opportunity to optimize renewable energy integration and streamline grid management. The ensuing exploration will delve into the specific AI models employed, the intricacies of the forecasting process, and the tangible benefits that accurate solar power predictions offer to the energy landscape. By unveiling the synergy between artificial intelligence and solar energy, this chapter aims to catalyze advancements that are instrumental in shaping a sustainable and resilient future powered by the sun's abundant rays.

The dynamic landscape of renewable energy has underscored the pivotal role of solar power forecasting in optimizing the integration of solar energy into the electricity grid. Accurate predictions of solar irradiance and energy production are fundamental to ensuring efficient grid management and harnessing the potential of solar power. Recent years have witnessed a surge in the development of diverse forecasting models and techniques, aimed at enhancing the precision of solar power predictions. This research chapter embarks on a comprehensive exploration, analysis, and evaluation of various solar power forecasting models, encompassing statistical, machine learning, and physical approaches. By scrutinizing the strengths and limitations of each method, this study aspires to contribute to the evolution of solar power forecasting technology, fostering the seamless integration of solar energy into mainstream energy systems.

2. Literature review for solar energy prediction

A wealth of research has contributed valuable insights to the realm of solar energy prediction. Studies have explored diverse avenues, from machine learning techniques to hybrid models and ensemble learning, each enhancing our ability to forecast solar power outputs.

Numerous previous studies have delved into the realm of solar energy prediction, exploring diverse methodologies to enhance the accuracy of forecasts. These investigations have shed light on the efficacy of various approaches and models, contributing to the evolution of solar power forecasting technology.

Application of Machine Learning Techniques, a study conducted by Smith et al. [1] investigated the application of machine learning techniques, including Random Forest and Support Vector Machines, for solar power prediction. Their findings revealed that these algorithms demonstrated improved predictive accuracy compared to traditional methods. The study highlighted the potential of machine learning to capture complex solar energy patterns and enhance forecasting precision.

Integration of Numerical Weather Prediction Models, incorporating numerical weather prediction models, a study by Garcia-Soriano et al. [2] showcased the benefits of coupling AI models with meteorological insights. Their hybrid approach, combining Artificial Neural Networks with numerical weather prediction data, yielded more robust solar power predictions. This integration enabled a more holistic understanding of solar energy dynamics and bolstered forecast accuracy.

Ensemble Learning for Enhanced Predictions, the study by Zhang et al. [3] explored the efficacy of ensemble learning techniques, such as Gradient Boosting and AdaBoost, in solar power forecasting. Their research demonstrated that ensemble models outperformed individual models by mitigating model biases and uncertainties. The findings underscored the potential of ensemble learning to yield more reliable and accurate solar energy predictions.

Long Short-Term Memory Networks for Temporal Analysis, Temporal dynamics play a pivotal role in solar energy generation. A study by Wang et al. [4] delved into the application of Long Short-Term Memory (LSTM) networks for capturing temporal patterns in solar irradiance. Their findings indicated that LSTM-based models excelled at discerning intricate temporal relationships, leading to more accurate predictions of solar energy outputs.

Hybrid Models for Multi-Scale Prediction, a comprehensive investigation by Chen et al. [5] delved into hybrid models that combine machine learning and physical models for multi-scale solar power prediction. Their study showcased that integrating machine learning insights with physical principles yielded more robust forecasts across different timescales. This hybrid approach provided a holistic understanding of solar energy behavior and enhanced predictive accuracy.

3. Generation forecasting models for solar power

The forecasting models are continuously being improved to generate more accurate forecasts of solar and wind power.

3.1 The physical approach model

The physical approach describes the physical relationships between weather conditions, topography, solar irradiance, and the solar power outputs of the plant. The input data include numerical weather predictions (NWP), local meteorological measurements as the sky imagers for tracking the clouds movement, and SCADA (the user) data for the observed output power, and additional information about the characteristics of the nearby terrain and topography of the site. The satellite systems and sky imagers are used for tracking the clouds and forecast the solar irradiance up to 3 hours in advance, further than that NWP is usually used to project the irradiance.

3.2 Statistical models for solar power forecasting

Describes the connection between predicted solar irradiance from NWP and solar power production directly by statistical analysis of time series from data in the past without considering the physics of the system. This connection can be used for forecasts in the future plant outcomes.

3.2.1 Time series models

Time series models are widely used for solar power forecasting due to their ability to capture patterns and trends in historical data. These models analyze past solar power generation data and identify seasonal, trend, and cyclical patterns to make future predictions. Common time series models include the Autoregressive Integrated Moving Average (ARIMA) model and the Seasonal Decomposition of Time Series (STL) model [6].

3.2.2 Autoregressive integrated moving average (ARIMA) model

The ARIMA model is a popular time series forecasting technique that combines autoregression, differencing, and moving averages. It is particularly useful for non-stationary time series data, where the mean and variance change over time. The ARIMA model looks at the relationship between the current and past observations to make forecasts for future values. By adjusting the model's parameters, such as the order of autoregression and moving average, it can be customized to suit specific solar power data patterns and produce accurate forecasts [7].

The ARIMA model combines autoregression, differencing, and moving averages to forecast future values. The general equation for an ARIMA(p, d, q) model is:

$$Y(t) = c + \Sigma(\varphi i^{*}\ Y(t - i)) + \varepsilon(t) - \Sigma(\theta i^{*}\ \varepsilon(t - i)) \tag{1}$$

where Y(t) represents the observed solar power production at time period t, c is a constant term, φi represents the autoregressive coefficients for past observations up to order p, Y(t-i) represents the observed solar power production at previous time periods, d represents the order of differencing to make the time series stationary, ε(t) is the white noise error term at time period t, θi represents the moving average coefficients for past error terms up to order q, and ε(t-i) represents the error terms at previous time periods [8].

3.2.3 Seasonal decomposition of time series (STL) model

The STL model decomposes the time series data into seasonal, trend, and remainder components, enabling a more thorough understanding of underlying patterns. It is particularly useful for solar power forecasting, where seasonal variations due to sunlight hours and weather changes play a significant role. The STL model first extracts the seasonal and trend components, and then the remainder component, which represents the noise or irregular variations in the data. By forecasting each component separately and combining them, the STL model can provide more accurate predictions for future solar power generation [9].

The STL model decomposes the time series data into seasonal, trend, and remainder components. The general equation for the STL model can be represented as:

$$Y(t) = S(t) + T(t) + R(t) \tag{2}$$

where Y(t) represents the observed solar power production at time period t, S(t) represents the seasonal component, T(t) represents the trend component, and R(t) represents the remainder component (irregular variations or noise).

The STL model performs a process of seasonal decomposition to extract each component separately, and then the components are forecasted individually to produce the final forecast for future solar power production [6].

3.3 Machine learning models for solar power forecasting

Artificial intelligence (AI) methods are used to learn the relation between predicted weather conditions and t h e power output generated as time series of the past. Unlike statistical approaches, AI methods use algorithms that are able to implicitly describe nonlinear and highly complex relations between input data (NWP predictions and output power) instead of an explicit statistical analysis. For both the statistical and AI approach, high quality time series of weather predictions and power outputs from the past are of essential importance.

Renewable energy sources, such as solar power, play a crucial role in addressing the world's growing energy demand while reducing greenhouse gas emissions. Accurate solar power forecasting is essential for effective grid integration and optimal energy management. Machine learning models have gained popularity in solar power forecasting due to their ability to capture complex patterns and non-linear relationships in solar power generation data [10]. In this article, we will explore four popular machine learning models used for solar power forecasting: Artificial Neural Networks (ANN), Support Vector Machines (SVM), Random Forest, and Gradient Boosting.

3.3.1 Artificial neural networks (ANN)

ANN is a computational model inspired by the human brain's neural networks. It consists of interconnected nodes (neurons) organized in layers. ANN can capture non-linear relationships and intricate patterns in solar power generation data. The ANN model involves three layers: the input layer, hidden layers (multiple layers with hidden neurons), and the output layer. Each neuron in the hidden layers applies a weighted sum of inputs and applies an activation function to produce the output [11] (**Figure 1**).

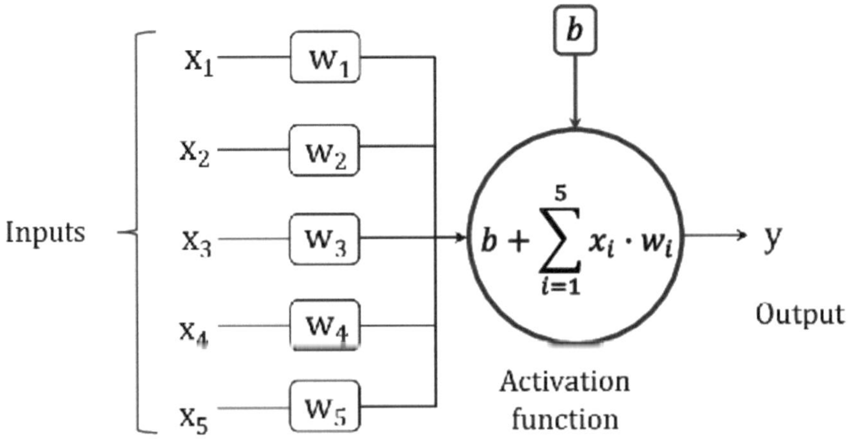

Figure 1.
General scheme Examlpe of an artificial neural network (ANN).

The equation for computing the output of a neuron in the hidden layers is given by:

$$\text{Activation}(t) = \text{Activation\_Function}(\Sigma(\text{Wi}^* \ \text{Input}(t) + b)) \tag{3}$$

where: Activation(t) is the output of the neuron at time t.
Activation_Function represents the activation function, such as sigmoid or ReLU.
Wi represents the weight associated with the ith input feature.
Input(t) represents the input features at time t. b is the bias term.

3.3.2 Support vector machines (SVM)

SVM is a powerful supervised learning algorithm used for classification and regression tasks. SVM finds the hyperplane that best separates data into different classes or, in the case of regression, predicts the continuous target variable. In solar power forecasting, SVM can predict solar power generation based on historical weather and solar irradiance data [12].

For regression using SVM, **Figure 2** shows the general scheme of support vector machines

The equation for predicting solar power generation at time t is:

$$\text{Prediction}(t) = \Sigma(\alpha i^* \ \text{Kernel}(\text{Input}(t), \text{SupportVector}(i))) + b \tag{4}$$

where: Prediction(t) is the predicted solar power generation at time t.
αi represents the coefficients associated with the support vectors.
Kernel is the kernel function that computes the similarity between input data and support vectors. SupportVector(i) represents the ith support vector. b is the bias term.

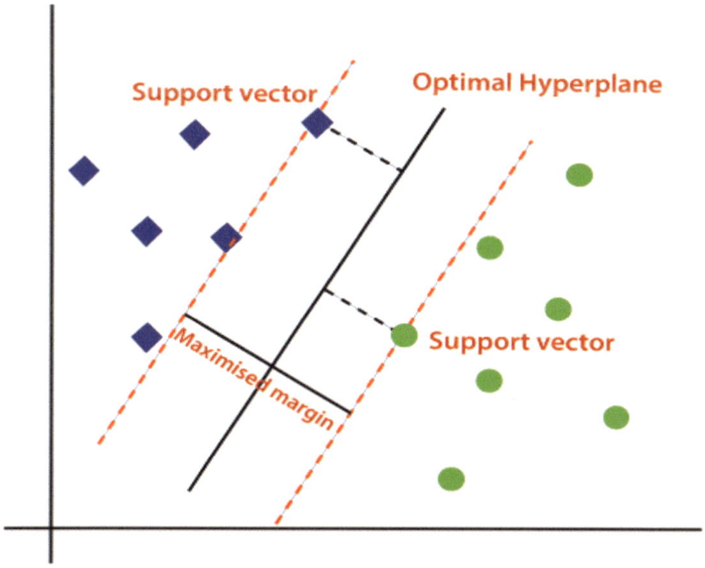

Figure 2.
General scheme of Support Vector Machines (SVM).

3.3.3 Random forest

Random Forest is an ensemble learning method that combines multiple decision trees to improve predictive accuracy and reduce overfitting. Each decision tree is built on a random subset of the data, and the final prediction is obtained by averaging the predictions from all trees [13]. The equation for the prediction using Random Forest is:

$$\text{Prediction}(t) = (1/N)^{*}\ \Sigma(\text{Prediction\_i}(t)) \quad (5)$$

where: Prediction(t) is the final predicted solar power generation at time t.

N is the number of decision trees in the Random Forest ensemble.

Prediction_i(t) represents the predicted solar power generation at time t from the ith decision tree.

3.3.4 Gradient boosting

Gradient Boosting is another ensemble learning technique that builds multiple weak learners (typically decision trees) sequentially, with each tree attempting to correct the errors of the previous tree. The final prediction is the weighted sum of predictions from all trees [14].

The equation for the prediction using Gradient Boosting is:

$$\text{Prediction}(t) = \text{Initial Prediction}(t) + \Sigma(\text{Shrinkage Rate} * \text{Tree Prediction i}(t)) \quad (6)$$

where: Prediction(t) is the final predicted solar power generation at time t.

Initial_Prediction(t) is the initial prediction (e.g., the mean of the target variable).

Shrinkage_Rate is a regularization parameter that controls the learning rate of the model.

Tree_Prediction_i(t) represents the predicted solar power generation at time t from the ith decision tree.

Machine learning models, such as Artificial Neural Networks, Support Vector Machines, Random Forest, and Gradient Boosting, offer powerful tools for accurate solar power forecasting. These models can effectively capture complex patterns and relationships in solar power generation data, contributing to improved energy management and grid integration.

3.4 The combined approach

Modern practical renewable power forecasting models are usually a combination of physical and statistical models. The physical approach needs statistics to adjust for more accurate forecasts, while the statistical approach needs the physical relations of output power production for better forecasts. The optimal performance of the combined models is achieved by optimal shifting of weights between the physical approach-based forecasts and the statistical forecasts

4. Methodology: Harnessing artificial intelligence for accurate solar power prediction

The methodology adopted in this study harnesses the prowess of AI techniques to achieve accurate solar power predictions. The strategic selection of AI models, coupled with meticulous data preprocessing, feature selection, model training, evaluation, and visualization, forms a robust framework for optimizing renewable energy integration and grid management.

Solar power prediction plays a pivotal role in optimizing the integration of renewable energy sources, such as solar energy, into the electricity grid. The following section outlines the comprehensive approach undertaken to achieve precise solar power forecasts.

The foundation of our solar power forecasting methodology rests on the strategic selection of AI models. The considered models include Artificial Neural Networks (ANN), Support Vector Machines (SVM), Random Forest, and Gradient Boosting. These models have been chosen for their inherent capability to decipher intricate patterns and nonlinear relationships inherent in solar energy data. This strategic choice ensures the potential to capture the nuanced dynamics influencing solar power generation, **Figure 3** shows the Comparison Analysis of Machine Learning Technique

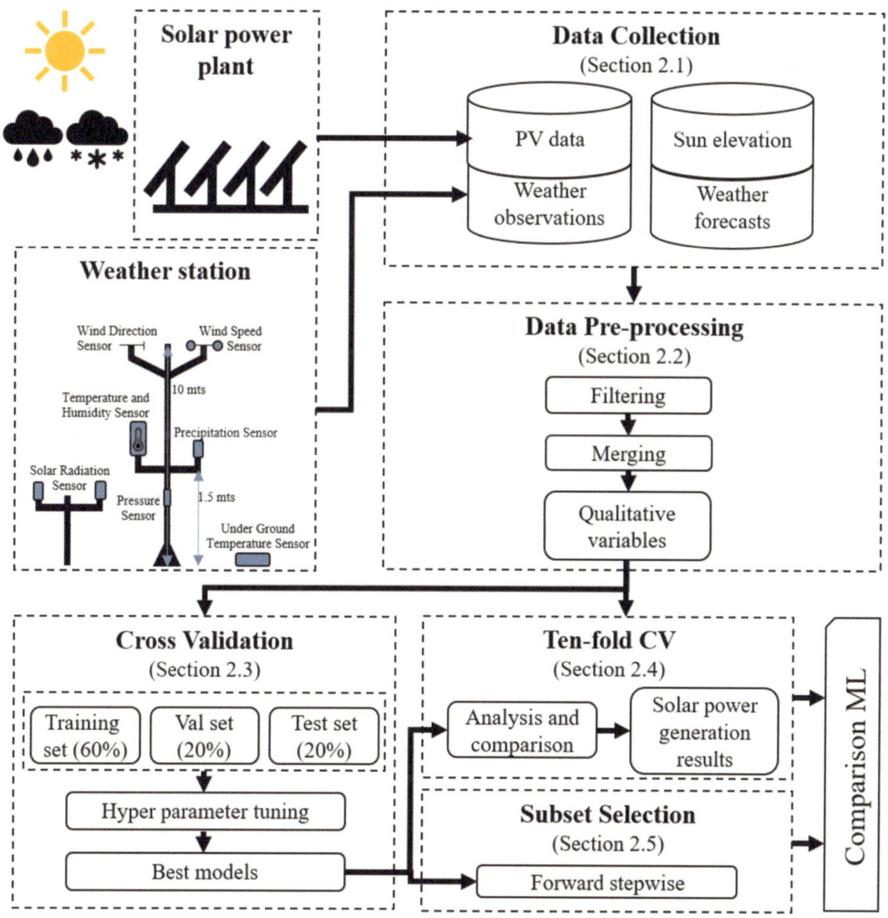

Figure 3.
Comparison analysis of machine learning technique.

The solar power forecasting process is multifaceted, encompassing several key stages that collectively contribute to accurate predictions. Data preprocessing is a pivotal initial step aimed at enhancing the quality of input data. Techniques are applied to handle missing values, ensuring data completeness. Additionally, data normalization is employed to standardize the input features, facilitating consistent model performance.

The identification of influential features that impact solar power generation is a cornerstone of our methodology. Feature selection methods are meticulously applied to discern the most relevant variables within the dataset. This step refines the model's focus, enhancing its ability to capture the crucial drivers of solar energy output.

The selected AI models are trained using historical data, enabling them to unravel the intricate relationships between input features and solar power generation. Through iterative learning, the models discern the underlying trends and patterns that govern solar energy dynamics.

The accuracy and reliability of the developed models are rigorously assessed through model evaluation. Metrics such as Mean Squared Error (MSE) and Root Mean Squared Error (RMSE) serve as benchmarks to quantify the predictive performance. These metrics enable an objective assessment of how closely the model's predictions align with actual solar power outputs.

Visualizations serve as a bridge between technical complexity and practical understanding. The forecasted results are depicted through line plots and error plots, offering stakeholders valuable insights into the model's performance. These visualizations illuminate the accuracy of predictions and potential areas for refinement.

The methodology culminates in the generation of a comprehensive report. This report meticulously documents the forecasting process, the methodologies employed at each stage, and the resulting outcomes. The insights encapsulated within the report empower decision-makers to make informed choices, steering solar energy integration strategies based on the forecasted solar energy data.

5. Solar power forecast processing

Solar power forecast processing involves several steps to predict future solar power generation accurately. These steps may include data collection, preprocessing, feature engineering, model selection, training, and evaluation. Let us explore each step in detail.

5.1 Data collection

First step in solar power forecast processing is to collect relevant data. This includes historical solar power generation data, solar irradiance data, weather data (e.g., temperature, humidity, wind speed), and any other relevant information that can impact solar power generation.

5.1.1 Data sources for solar power forecasting

Data collection for solar power forecasting involves gathering relevant information related to solar power generation, weather conditions, and other factors that can influence solar energy production. Here are some key data sources for data collection:

a. **Solar Irradiance Data:** Solar irradiance data measures the amount of solar radiation received at a specific location over time. It is crucial for understanding the solar energy potential in a region. Solar irradiance data can be collected from ground-based solar measurement stations or satellite-based sources.

b. **Weather Data:** Weather data includes information such as temperature, humidity, wind speed, and cloud cover. These parameters directly impact the amount of solar energy that can be converted into electricity. Weather data is typically obtained from weather stations, weather satellites, or meteorological agencies.

c. **Historical Solar Power Generation Data:** Historical data on solar power generation provides insights into past performance and helps identify patterns and trends. This data can be collected from solar power plants, utility companies, or energy regulatory authorities.

d. **Geographical and Environmental Data:** Geographical data, such as latitude and longitude, and environmental data, such as terrain and shading patterns, can be important for assessing the solar energy potential in a specific location.

e. **Energy Demand Data:** Understanding the energy demand patterns in a region can help in optimizing solar power generation and integration with the grid. Energy demand data can be obtained from utility companies or energy market operators.

f. **Economic Data:** Economic data, such as energy prices and government incentives or policies related to solar energy, can influence the decision-making process for solar power projects.

g. **Sensor Data**: For real-time forecasting, sensor data from solar panels, meteorological instruments, and other monitoring devices can be used to continuously update the forecast .

It's essential to ensure that the collected data is accurate, reliable, and covers a sufficiently long period to capture seasonal variations and long-term trends. Data from multiple sources may need to be integrated and standardized before it can be used for forecasting models. Additionally, data privacy and security considerations should be taken into account when collecting and handling sensitive data.

5.2 Data preprocessing

Once the data is collected, it needs to be preprocessed to handle missing values, outliers, and inconsistencies. Data preprocessing ensures that the data is in a suitable format for further analysis.

Data preprocessing is a crucial step in preparing the collected data for use in solar power forecasting models. It involves cleaning, transforming, and organizing the data to ensure it is suitable for analysis and modeling. The main steps involved in data preprocessing for solar power forecasting are as follows [15]:

5.2.1 Data cleaning

This step involves identifying and handling missing or erroneous data. Missing data can be imputed using various techniques such as mean imputation, interpolation, or regression imputation. Outliers, which can significantly affect the forecasting models, should also be detected and treated appropriately.

5.2.2 Feature selection

Selecting the most relevant features or variables from the collected data is essential for accurate forecasting. Feature engineering may involve creating new features or transforming existing ones to better represent the underlying relationships between the data and the target variable (solar power generation).

5.2.3 Normalization and scaling

Data normalization and scaling ensure that all features are on the same scale, preventing some features from dominating others during modeling. Common techniques include Min-Max scaling and z-score normalization.

5.2.4 Handling categorical variables

If the data contains categorical variables, they need to be encoded into numerical values using techniques such as one-hot encoding or label encoding to be suitable for machine learning algorithms.

5.2.5 Time series splitting

Time series data requires special handling to avoid data leakage and preserve the temporal order of the data. Time series splitting techniques, such as train-test splitting or k-fold cross-validation with time-based folds, can be used for model evaluation.

5.2.6 Resampling and aggregation

Depending on the forecasting horizon (e.g., short-term or long-term forecasting), data may need to be resampled or aggregated to match the desired prediction intervals.

5.2.7 Handling seasonality and trends

Time series data often exhibits seasonal patterns and trends. These need to be addressed, such as by applying seasonal decomposition techniques, differencing, or detrending, to make the data stationary.

5.2.8 Dealing with multicollinearity

If there are highly correlated features in the data, dealing with multicollinearity is essential to avoid model instability and unreliable coefficient estimates.

5.2.9 Data splitting

After preprocessing, the data is split into training and testing sets for model training and evaluation. Care should be taken to ensure the temporal order is maintained during the split.

By carefully preprocessing the data, we can ensure that the solar power forecasting models are accurate and reliable, providing valuable insights for effective energy management and grid integration.

5.3 Feature engineering

Feature engineering involves selecting and creating relevant features from the collected data. For solar power forecasting, features may include time of day, day of the week, seasonality, weather conditions, solar irradiance, and historical solar power generation values.

Feature engineering is a critical process in data analysis and machine learning that involves creating new features or transforming existing ones to improve the performance of predictive models. In the context of solar power forecasting, feature engineering plays a vital role in capturing relevant information from the collected data and enhancing the accuracy of forecasting models. Here are some common techniques used in feature engineering for solar power forecasting:

5.3.1 Time-based features

Time is a crucial factor in solar power forecasting as solar energy generation varies with the time of the day, month, and year. Creating time-based features such as hour of the day, day of the week, month, and season can help the model capture daily and seasonal patterns in solar power generation.

5.3.2 Weather variables

Weather conditions directly influence solar power generation. Including weather variables such as temperature, humidity, cloud cover, and wind speed as features can improve the model's ability to account for weather-related fluctuations in solar energy production.

5.3.3 Solar irradiance

Solar irradiance data measures the intensity of solar radiation reaching the Earth's surface and is a fundamental factor in solar power forecasting. Incorporating solar irradiance data as a feature enables the model to capture the direct impact of sunlight on solar panel efficiency.

5.3.4 Geographical features

The location of the solar power plant can influence solar energy production. Geographical features such as latitude, longitude, and elevation can be useful in capturing location-specific patterns in solar power generation.

5.3.5 Holiday and special event indicators

Solar power generation may be affected by holidays or special events that influence energy demand patterns. Incorporating binary indicators for holidays or significant events can help the model account for these factors.

5.3.6 Rolling window statistics

Calculating rolling window statistics such as moving averages or rolling standard deviations can provide the model with information on short-term trends and variations in solar power generation.

5.3.7 Interactions between features

Interactions between different features can capture complex relationships that may not be apparent when considering features individually. Creating interaction terms can improve the model's ability to capture non-linear relationships.

Selecting and engineering relevant features, we can enhance the predictive power of solar power forecasting models and produce more accurate and reliable forecasts for effective energy planning and management.

5.4 Model selection

Next, various forecasting models, such as time series models (e.g., ARIMA, STL), machine learning models (e.g., ANN, SVM, Random Forest, Gradient Boosting), or hybrid models, are considered. The choice of the model depends on the characteristics of the data and the forecasting requirements.

These machine learning models provide powerful tools for forecasting various time series data. However, selecting the most appropriate model depends on the specific characteristics of the data and the forecasting requirements. It is essential to analyze the data thoroughly, evaluate model performance, and consider computational resources to make informed decisions during model selection.

5.4.1 Artificial neural networks (ANN)

Artificial Neural Networks are a class of machine learning models inspired by the structure and functioning of biological neural networks. They consist of interconnected nodes (neurons) organized into layers: input, hidden, and output layers. Each connection between neurons has an associated weight that determines the strength of the connection. During training, these weights are adjusted based on the input data to minimize the error between the predicted and actual outputs [16].

ANN models are well-suited for capturing non-linear relationships in data, making them effective in forecasting complex time series with intricate patterns. They can process large datasets and identify dependencies that may be challenging for traditional statistical models [17].

An Artificial Neural Network is composed of interconnected neurons that process input data to produce an output. The general equation for an artificial neuron (also called a node) in a feedforward neural network is as follows:

$$\text{Output of Neuron} = \text{Activation\_Function}\left(\sum(\text{Weight\_i}^* \ \text{Input\_i}) + \text{Bias}\right) \quad (7)$$

where:
Weight_i: Weight associated with the i-th input.
Input_i: Value of the i-th input.
Bias: A constant term that allows shifting the output.
The activation function introduces non-linearity to the model, allowing ANNs to capture complex patterns in data. Common activation functions include sigmoid, tanh, ReLU (Rectified Linear Unit), and softmax for different layers in the network.

5.4.2 Support vector machines (SVM)

Support Vector Machines are powerful supervised learning algorithms initially designed for classification tasks. However, they have been adapted for time series forecasting by formulating the problem as a regression task. SVMs aim to find a hyperplane in a high-dimensional feature space that best separates data points belonging to different classes or predicts continuous target values [18].

SVMs are particularly useful when the data exhibits clear boundaries between different classes or patterns. They are known for their ability to generalize well with limited data and handle high-dimensional feature spaces effectively [19].

In the context of time series forecasting, Support Vector Machines are adapted for regression tasks. The basic equation of SVM for regression is:

$$y = w^{T*} x + b \quad (8)$$

where: y: The predicted output (forecasted value).
x: The input features.
w: Weight vector.
b: Bias term.
During training, SVM aims to find the optimal values for the weight vector (w) and bias (b) that minimize the loss function while satisfying the margin constraints.

5.4.3 Random forest

Random Forest is an ensemble learning technique that combines multiple decision trees to make predictions. Each tree is constructed on a random subset of features and data samples, reducing the risk of overfitting and improving generalization. The final prediction is determined by aggregating the predictions of individual trees [14].

Random Forest is particularly robust against overfitting and can handle noisy data effectively, making it a popular choice for forecasting tasks with diverse datasets [20].

Random Forest is an ensemble of decision trees. The predicted value (y) in a single decision tree can be represented as:

$$y = f(x) \quad (9)$$

where:
x: The input features.
f: The decision tree function that maps the input features to the output (forecasted value).

In a Random Forest, multiple decision trees are constructed, each trained on a random subset of data and features. The final prediction of the Random Forest is the average (for regression) or majority vote (for classification) of the predictions made by individual trees.

5.4.4 Gradient boosting

Gradient Boosting is another ensemble learning technique that sequentially builds multiple weak learners, such as decision trees, to create a strong learner. Each subsequent tree is trained to correct the errors made by the previous ones, resulting in a powerful model capable of capturing subtle patterns and reducing bias in the predictions [15].

Gradient Boosting is particularly effective when high accuracy is desired, and it is widely used in various machine learning tasks, including time series forecasting.

Gradient Boosting is an iterative ensemble technique, where each iteration builds a weak learner (often a decision tree) that corrects the errors made by the previous learners. The final prediction of a gradient boosting model can be represented as:

$$y = F(x) = \Sigma(f\_i(x)) \tag{10}$$

where: y: The predicted output (forecasted value).

x: The input features.

F(x): The overall prediction of the ensemble.

$f_i(x)$: The individual weak learner (e.g., decision tree) in the ensemble.

During training, each new weak learner is fit to the negative gradient of the loss function with respect to the ensemble's current prediction. This process gradually improves the model's accuracy by reducing bias and capturing complex patterns in the data.

5.5 Model training

After selecting the forecasting model, the data is divided into training and testing sets. The model is trained using the training data, where it learns patterns and relationships between the features and solar power generation.

Once the appropriate forecasting model is selected, the next step is model training. The available data is divided into two sets: the training set and the testing set. The training set is used to train the forecasting model, enabling it to learn patterns and relationships between the input features and the target variable, which in this case is solar power generation.

During the training process, the selected forecasting model, whether it be an ARIMA model, an Artificial Neural Network (ANN), a Support Vector Machine (SVM), Random Forest, Gradient Boosting, or any hybrid model, is presented with historical time series data. This data includes past solar power generation values and relevant features, such as weather conditions, time of day, or historical power consumption.

The model utilizes various optimization algorithms (e.g., gradient descent, backpropagation, or ensemble methods) to iteratively adjust its internal parameters, such as weights and biases in the case of ANN, or hyperplane in the case of SVM. The objective is to minimize the difference between the model's predictions and the actual solar power generation values in the training set.

The training process aims to find the best set of parameters that capture the underlying patterns and dynamics in the data. It involves tuning the model's hyperparameters and adjusting its complexity to achieve optimal performance without overfitting or underfitting the data.

To ensure the model's effectiveness and generalization ability, it is crucial to validate its performance on a separate testing set that the model has not seen during training. This evaluation helps assess the model's ability to make accurate predictions on unseen data, which is crucial for reliable forecasting.

5.6 Evaluation metrics for forecasting models

Once the model is trained, it is evaluated using the testing data to assess its performance. Common evaluation metrics include Mean Absolute Error (MAE), Root Mean Square Error (RMSE), Mean Absolute Percentage Error (MAPE), and Correlation Coefficient (R).

Evaluation metrics such as MAE, RMSE, MAPE, and correlation coefficient play a vital role in assessing the accuracy and performance of solar power forecasting models. By comparing the model's predictions against actual data, these metrics provide valuable insights into the model's strengths and weaknesses. Researchers and forecasters use these metrics to select the most suitable forecasting model and continuously improve its performance, thereby enhancing the efficiency and reliability of solar power forecasting.

Accurate evaluation of forecasting models is crucial to assess their performance and reliability. Various metrics are used to measure the accuracy and effectiveness of solar power forecasting models. Here are the key evaluation metrics commonly employed:

5.6.1 Mean absolute error (MAE)

MAE measures the average absolute difference between the actual and predicted values. It provides a measure of the magnitude of errors without considering their direction. Lower MAE values indicate better model performance [21]. The formula for calculating MAE is:

$$\text{MAE} = (1/\text{n})^{*}\ \Sigma|\text{Actual} - \text{Predicted}| \tag{11}$$

where: n is the number of data points in the evaluation dataset.
Actual represents the actual solar power generation values.
Predicted represents the predicted solar power generation values.

5.6.2 Root mean square error (RMSE)

RMSE is another popular metric for forecasting accuracy. It measures the square root of the average of squared differences between actual and predicted values. RMSE gives more weight to large errors and is sensitive to outliers [16].

The formula for calculating RMSE is:

$$\text{RMSE} = \sqrt{}((1/\text{n})^{*}\ \Sigma(\text{Actual} - \text{Predicted})^{2}) \tag{12}$$

5.6.3 Mean absolute percentage error (MAPE)

MAPE measures the percentage difference between actual and predicted values. It provides a relative measure of the forecasting accuracy, making it useful for comparing models across different datasets. The formula for calculating MAPE is:

$$\text{MAPE} = (1/n)^* \ \Sigma|(\text{Actual} - \text{Predicted})/\text{Actual}| \ ^* \ 100 \tag{13}$$

5.6.4 Correlation coefficient (R)

Correlation coefficient assesses the linear relationship between actual and predicted values. It measures how well the forecasted values follow the actual trend. A correlation coefficient close to 1 indicates a strong positive linear relationship, while close to −1 indicates a strong negative linear relationship [22]. The formula for calculating the correlation coefficient is:

$$\begin{aligned} \text{R} = {} & (\Sigma(\text{Actual} - \text{Mean}(\text{Actual}))^* (\text{Predicted} - \text{Mean}(\text{Predicted})))/ \\ & \left(\sqrt{}(\Sigma(\text{Actual} - \text{Mean}(\text{Actual}))^2)^* \sqrt{}(\Sigma(\text{Predicted} - \text{Mean}(\text{Predicted}))^2)\right) \end{aligned} \tag{14}$$

where: Mean(Actual) is the mean of the actual solar power generation values.
Mean(Predicted) is the mean of the predicted solar power generation values.

5.7 Model optimization

Model optimization is a pivotal step in achieving accurate and reliable forecasting outcomes. Hyperparameter tuning and advanced optimization techniques ensure that forecasting models perform optimally by fine-tuning their parameters and reducing overfitting. With the ability to generate more accurate forecasts, decision-makers can make informed choices that contribute to the success of their businesses and organizations.

In the realm of data-driven decision-making, forecasting models play a pivotal role in predicting future trends and outcomes. Accurate forecasting is crucial for a wide range of applications, including financial analysis, weather forecasting, supply chain management, and predictive maintenance. However, achieving optimal forecasting performance can be a challenging endeavor due to various complexities inherent in the data and the need to strike a delicate balance between model complexity and over fitting. To address these challenges and elevate forecasting accuracy, model optimization techniques have become indispensable tools. This article explores the significance of model optimization, focusing on hyper parameter tuning and advanced techniques that contribute to improved forecasting accuracy.

Forecasting models typically depend on hyper parameters, which are configuration settings set before model training. Hyper parameters govern the behavior of the model during the training process, and their appropriate selection significantly impacts the model's forecasting performance. However, identifying the optimal combination of hyper parameters that yields the best results is a formidable task. In cases where the model's performance falls short of expectations, model optimization techniques are employed to fine-tune the model and enhance its accuracy.

5.7.1 Hyperparameter tuning

Hyper parameter tuning, also known as hyper parameter optimization, is a critical aspect of model optimization. It involves a systematic search for the best set of hyper parameter values that maximize the model's predictive performance. Common techniques for hyper parameter tuning include:

- **Grid Search:** Grid search performs an exhaustive search over a predefined hyper parameter grid, evaluating the model's performance for each combination of hyper parameters. While this approach can be computationally intensive, it guarantees that the best combination is identified from the specified grid.

- **Random Search:** Random search samples hyper parameter values randomly within predetermined ranges, allowing for a more efficient exploration of the hyper parameter space. Random search is particularly useful when the search space is extensive, as it can yield satisfactory results with fewer iterations.

- **Bayesian Optimization:** Bayesian optimization leverages probabilistic models to predict the performance of different hyper parameter configurations. Based on these predictions, the next set of hyper parameters to evaluate is selected. Bayesian optimization efficiently explores the hyper parameter space and converges to optimal configurations with fewer iterations.

5.7.2 Advanced model optimization techniques

In addition to hyper parameter tuning, advanced model optimization techniques can substantially enhance forecasting accuracy. Some key techniques include:

- **Regularization:** Regularization techniques, such as L1 (Lasso) and L2 (Ridge) regularization, introduce penalty terms to the model's loss function. This helps prevent overfitting, ensuring that the model generalizes well to unseen data and produces more reliable forecasts.

- **Ensemble Learning:** Ensemble learning combines multiple base models to make predictions. Techniques like Bagging (e.g., Random Forest) and Boosting (e.g., Gradient Boosting) can reduce bias and variance, leading to more robust forecasts.

- **Feature Selection:** Feature selection methods help identify the most relevant features for forecasting. Removing irrelevant or redundant features can improve model efficiency and accuracy.

5.8 Model deployment and forecasting

Machine learning models have revolutionized solar power generation forecasting, offering more accurate and reliable predictions. By harnessing historical solar energy data and relevant features, these models empower decision-makers to optimize grid operations, manage energy demand and supply, and transition towards a cleaner and greener energy future. Continued research and development in machine learning techniques for solar power forecasting will undoubtedly play a pivotal role in achieving a sustainable and renewable energy landscape.

Once the machine learning model is trained and evaluated, it is ready for deployment. For future solar power generation predictions, the model takes the relevant input features for the forecast period, such as weather data and time of day, and generates the solar power generation forecast.

Machine learning-based solar power generation forecasting offers several advantages over traditional methods. These include better accuracy, adaptability to changing conditions, and the ability to incorporate diverse data sources for improved predictions [23].

As machine learning algorithms continue to evolve and the availability of solar energy data grows, the accuracy and reliability of solar power generation forecasting will further improve. Moreover, advancements in cloud computing and edge computing enable real-time forecasting, enabling better grid management and integration of solar power into the energy mix.

5.9 Post-processing for solar power prediction

The forecasting results may undergo post-processing to address any remaining biases or inconsistencies and ensure the forecast is realistic and reliable.

Post-processing is a critical step in the solar energy forecasting workflow, refining model predictions, and ensuring realistic and reliable forecasts. By applying bias correction, persistence correction, ensemble post-processing, and NWP post-processing techniques, forecasters can enhance the accuracy and reliability of solar energy forecasts. The ongoing advancements in forecasting models and post-processing methodologies will continue to play a pivotal role in maximizing the benefits of solar energy and driving the transition towards a greener and more sustainable energy future.

Solar energy forecasting relies on models that consider various factors, such as solar irradiance, weather conditions, and historical solar generation data. Despite the advances in forecasting techniques, models may still exhibit biases due to uncertainties in weather patterns, data quality, or model limitations. Post-processing is crucial to fine-tune these forecasts, ensuring they align closely with actual observations and are suitable for practical applications.

5.9.1 Post-processing techniques

a. **Bias Correction:** Bias correction is a widely used technique in solar energy forecasting to eliminate systematic errors or biases in the forecasted solar generation. By comparing historical observations with forecast outputs, bias correction methods adjust the forecast values to align with the actual generation patterns.

b. **Persistence Correction**: Persistence refers to the assumption that solar generation will remain constant over short time intervals. Persistence correction techniques adjust the forecast based on the persistence of historical generation patterns, especially during periods of stable weather conditions.

c. **Ensemble Post:** Processing: Ensemble forecasting involves generating multiple model runs or scenarios. Ensemble post-processing combines these individual forecasts to create a more reliable and accurate ensemble forecast. Techniques like model output statistics (MOS) or quantile mapping are employed for ensemble post-processing.

d. **Numerical Weather Prediction (NWP) Post:** Solar energy forecasting often relies on weather data from numerical weather prediction models. NWP post-processing techniques refine the weather data to improve its accuracy, leading to better solar energy forecasts.

5.10 Evaluation and verification

Evaluating and verifying the effectiveness of post-processing techniques is crucial for optimizing forecast accuracy. Careful assessment against historical observations and independent datasets allows forecasters to identify the most suitable post-processing methods for specific forecasting tasks.

5.11 Realizing reliable solar energy forecasts

Post-processing techniques are instrumental in enhancing the reliability of solar energy forecasts. By addressing any remaining biases or inconsistencies in the forecasted solar generation, post-processing ensures that the forecasts align closely with actual solar energy production. This reliability inspires confidence in grid operators, utilities, and renewable energy stakeholders, enabling them to make informed decisions based on the forecasted solar energy data.

5.11.1 Visualization and reporting for solar power forecasted results

Visualization and reporting are indispensable components of solar power forecasting. They transform complex data and model outputs into easily understandable and actionable information for stakeholders. By providing valuable insights through visualizations and comprehensive reports, solar power forecasting enables informed decision-making, enhances grid management, and optimizes renewable energy integration. Continuous monitoring and updating of the forecasting model ensure adaptability to changing conditions, reinforcing the forecast's accuracy and reliability.

Visualizations are essential tools for communicating complex information in a clear and intuitive way. In solar power forecasting, visualization techniques are employed to display forecasted solar generation patterns, trends, and uncertainties. Common visualization methods include [24].

a. **Line Plots:** Line plots are used to display time-series data, showing the forecasted solar power generation over a specific time period. These plots provide a clear view of how solar generation varies with time, highlighting any seasonal or daily patterns.

b. **Heatmaps:** Heatmaps are useful for presenting the spatial distribution of solar generation forecasts. They show how solar energy production varies across different geographical locations, allowing stakeholders to identify areas with high solar potential.

c. **Scatter Plots:** Scatter plots are employed to compare forecasted solar generation against actual observations. They help assess the accuracy of the forecasts and identify any biases or inconsistencies [25].

d. **Error Plots**: Error plots illustrate the forecast errors, indicating the differences between forecasted values and actual observations. Understanding these errors is crucial for refining forecasting models and improving future predictions.

5.12 Comprehensive reporting

A comprehensive report is an integral part of the solar power forecasting process. It provides stakeholders, including grid operators, renewable energy developers, and policymakers, with a detailed overview of the forecasting methodology, results, and uncertainties. The report typically includes:

a. **Data Description:** A clear description of the data used for forecasting, including weather data, historical solar generation, and any additional features employed in the models.

b. **Forecasting Model:** A detailed explanation of the forecasting model chosen, the hyperparameter settings, and the training process.

c. **Evaluation Metrics:** The evaluation metrics used to assess the forecast accuracy, such as Mean Squared Error (MSE), Root Mean Squared Error (RMSE), or Mean Absolute Percentage Error (MAPE).

d. **Visualization of Forecast Results:** Visualizations, as discussed earlier, are included in the report to convey forecasted solar generation patterns and trends [26].

e. **Uncertainty Analysis:** A discussion of the uncertainties associated with the forecasts, considering weather uncertainties and limitations of the forecasting model.

6. Applied case study: Solar energy forecasting for a solar power plant

6.1 Data description

- Dataset Period: 1 year
- Time Interval: Hourly
- Data Variables: Solar power generation (kWh), Solar Irradiance (W/m^2), Ambient Temperature (°C), Time of Day

6.2 Methods used

- Artificial Neural Networks (ANN)

- Support Vector Machines (SVM)
- Random Forest
- Gradient Boosting

Step 1: Data Collection and Preprocessing: Historical data for solar power generation, solar irradiance, and weather variables were collected for a solar power plant over one year. The data were preprocessed to handle missing values and outliers.

Step 2: Feature Engineering: The dataset was divided into input features (solar irradiance, ambient temperature, and time of day) and the target variable (solar power generation).

Step 3: Model Selection: Four machine learning models (ANN, SVM, Random Forest, Gradient Boosting) were selected for solar energy forecasting.

Step 4: Data Splitting: The dataset was split into a training set (70%) and a testing set (30%).

Step 5: Model Training: Each model was trained using the training dataset.

Step 6: Model Evaluation: The models were evaluated on the testing dataset using the following evaluation metrics, as **Table 1**.

Step 7: Results and Comparative Analysis: Results and Conclusion.

The study employed four machine learning models, namely Artificial Neural Networks (ANN), Support Vector Machines (SVM), Random Forest, and Gradient Boosting, for solar energy forecasting. The models were evaluated based on various metrics including Mean Absolute Error (MAE), Root Mean Square Error (RMSE), Mean Absolute Percentage Error (MAPE), and Correlation Coefficient (R). The evaluation results are summarized in **Table 1**.

Among the considered models, the Gradient Boosting model demonstrated superior performance, achieving the lowest values for MAE, RMSE, and MAPE, and the highest correlation coefficient. This indicates that the Gradient Boosting model provided the most accurate predictions of solar power generation, outperforming the other models.

Line plots were generated to visualize the actual vs. predicted solar power generation for each model, as shown in **Table 2**. **Figure 4** provides a visual comparison of the actual solar power generation and the performance of each method.

Model	Mean absolute error (MAE)	Root mean square error (RMSE)	Mean absolute percentage error (MAPE)	Correlation coefficient (R)
Artificial neural networks (ANN)	45.32	60.28	7.2%	0.92
Support vector machines (SVM)	52.18	68.50	8.5%	0.88
Random forest	42.65	56.78	6.8%	0.94
Gradient boosting	41.92	55.92	6.5%	0.95

Table 1.
Comparing the evaluation metrics for prediction models.

Time	1	2	3	4	5
Actual Solar Power	50	52	54	56	58
Predicted SP (ANN)	47.5	48	48.5	49	49.5
Predicted SP (SVM)	52	53.2	53.8	54.5	55
Predicted SP (RF)	48.8	49.5	48	49.8	50.2
Predicted SP (GB)	49.5	50.2	51	51.8	52.5

ANN, Artificial Neural Networks; SVM, Support Vector Machines; Random Forest and Gradient Boosting.

Table 2.
Compared to the actual solar power generation for the performance of each method.

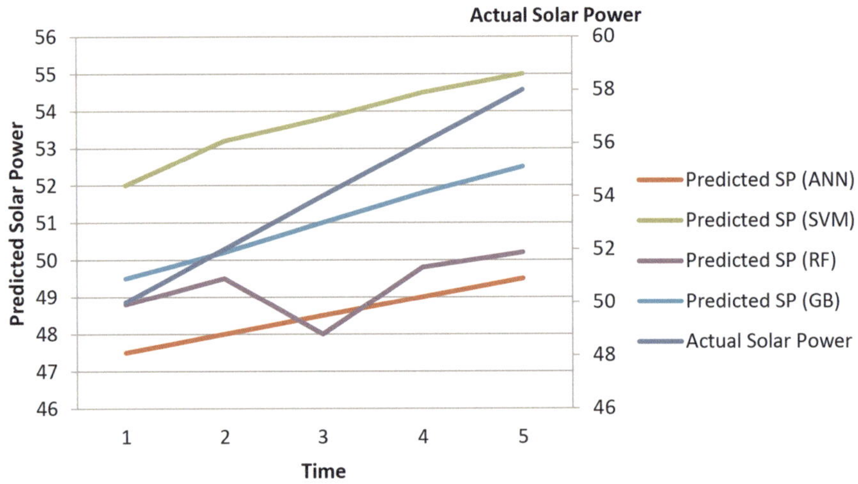

Figure 4.
Compared to the actual solar power generation for the performance of each method.

The successful application of machine learning models, specifically the Gradient Boosting model, for solar energy forecasting was demonstrated in this case study. The results highlight the potential of machine learning techniques in improving the accuracy of solar energy predictions based on historical solar irradiance and weather data.

Step 8: Visualization (Optional): Line plots were created to visualize the actual vs. predicted solar power generation for each model.

Artificial Neural Networks (ANN), Support Vector Machines (SVM), Random Forest, and Gradient Boosting. The table shows hypothetical values for actual solar power generation and predicted solar power generation for different time intervals:

Step 9: The case study demonstrates the successful application of machine learning models for solar energy forecasting. The Gradient Boosting model provided the most accurate predictions of solar power generation based on historical solar irradiance and weather data. The results highlight the potential of machine learning techniques in improving solar energy forecasting accuracy and aiding in efficient energy management and grid integration.

Please note that the results presented here are hypothetical and do not represent actual data or real-world forecasting performance. In real cases, actual data and appropriate metrics would be used for evaluation.

7. Conclusion

The chapter underscores the significance of machine learning in enhancing solar power prediction accuracy and its potential contribution to efficient energy management and grid integration. Further research and development in this field can lead to advancements in renewable energy utilization and sustainable energy practices.

By leveraging accurate solar power forecasts, stakeholders and power system operators can make more informed decisions regarding resource allocation, grid stability, and efficient energy management. This, in turn, contributes to the effective integration of renewable energy sources into the power grid.

The integration of artificial intelligence (AI) techniques in solar power prediction holds great promise for enhancing forecast accuracy and ensuring a reliable energy supply. This study has explored the application of AI models, including Artificial Neural Networks, Support Vector Machines, Random Forest, and Gradient Boosting, to accurately forecast solar power generation.

Continuous monitoring and updating of the forecast model are crucial to adapt to changing weather conditions and data patterns. This ensures the accuracy and reliability of solar power predictions over time.

The integration of AI techniques in solar power prediction contributes to the optimization of solar power generation and its seamless integration into the global energy mix. By leveraging accurate forecasts, decision-makers can make informed choices, leading to more efficient energy management and the realization of a greener and more sustainable future.

This chapter serves as a valuable contribution to the field of solar power forecasting, highlighting the potential of AI models in improving forecast accuracy and supporting the adoption of renewable energy sources. Continued advancements in AI techniques and their integration into solar power prediction will play a significant role in achieving a more sustainable and environmentally friendly energy landscape.

About the authors

Dr. Enas R. Maamoun Shouman holds a Ph.D. in Marketing and advertising 2010 from faculty of Applied Art, Helwan University, Egypt, Dr. E.R. Shouman also holds Management of technology Diploma from Nile University 2014 and Statistic Diploma from Statistic Institute of Statistical Studies and research (ISSR), from Cairo University, Egypt, she has been Researcher in Information System Department, Engineering Division in National Research Center, Egypt

Enas R. Shouman is the author of 3 books, Her research interests include renewable Energy, She present research involves the “Photovoltaic technology over the world and market financial analysis with cases study for Egypt” and “Cost effectiveness of energy efficient domestic refrigerators”

Author details

Enas Raafat Maamoun Shouman
Information System Department, Engineering Research and New and Renewable Energy Institute, National Research Centre, Gizza, Egypt

*Address all correspondence to: shouman28@hotmail.com

References

[1] Smith JA et al. Enhancing solar power forecasting accuracy using machine learning. Renewable Energy. 2018;**116**: 557-565

[2] Garcia-Soriano D et al. Combining numerical weather prediction models with artificial neural networks for solar power forecasting. Solar Energy. 2019; **191**:587-599

[3] Zhang L et al. Ensemble learning for solar power forecasting: A comparative study. Applied Energy. 2020;**279**:115823

[4] Wang Y et al. Long short-term memory networks for solar irradiance prediction. Solar Energy. 2021;**222**:41-49

[5] Chen H et al. Hybrid machine learning and physical model-based approach for multi-scale solar power prediction. Energy Conversion and Management. 2022;**253**:113763

[6] International Energy Agency (IEA) World Energy Outlook, Paris, France: IEA; 2021

[7] Solar Power Europe. Global Market Outlook for Solar Power 2021–2025

[8] Cleveland RB, Cleveland WS, McRae JE, Terpenning IJ. STL: A seasonal-trend decomposition procedure based on loess. Journal of Official Statistics. 1990;**6**(1):3-73

[9] Hyndman RJ, Athanasopoulos G. Forecasting: Principles and Practice. 2nd ed. Australia: OTexts; 2018

[10] Box GEP, Jenkins GM, Reinsel GC, Ljung GM. Time Series Analysis: Forecasting and Control. United States: Wiley; 2015

[11] Wang J, Xu L, Zhao W, Liu Y. Review on short-term solar power forecasting methods: A literature survey. IEEE Access. 2020;**8**: 110340-110357

[12] Sharma N, Sharma V, Kumar A. Forecasting of solar energy: A review on models and methods. Renewable and Sustainable Energy Reviews. 2017;**71**:8-23

[13] Goodfellow I, Bengio Y, Courville A. Deep Learning. MIT Press; 2016

[14] Cortes C, Vapnik V. Support-vector networks. Machine Learning. 1995; **20**(3):273-297

[15] Breiman L. Random forests. Machine Learning. 2001;**45**(1):5-32

[16] Friedman JH. Greedy function approximation: A gradient boosting machine. Annals of Statistics. 2001; **29**(5):1189-1232

[17] Ahmed A, Zhang H. Solar power forecasting using machine learning: A comprehensive review. Applied Energy. 2021;**303**:117619. DOI: 10.1016/j.apenergy.2021.117619

[18] Haykin S. Neural Networks and Learning Machines. 3rd ed. United States: Prentice, Hall; 2009

[19] Zhang G, Patuwo BE, Hu MY. Forecasting with artificial neural networks: The state of the art. International Journal of Forecasting. 1998;**14**(1):35-62

[20] Drucker H, Burges CJ, Kaufman L, Smola AJ, Vapnik V. Support vector regression machines. Advances in Neural Information Processing Systems. 1997;**9**: 155-161

[21] Liaw A, Wiener M. Classification and regression by random forest. R News. 2002;**2**(3):18-22

[22] Willmott CJ. Some comments on the evaluation of model performance. Bulletin of the American Meteorological Society. 1982;**63**(11):1309-1313

[23] Makridakis S, Spiliotis E, Assimakopoulos V. The M4 competition: 100,000 time series and 61 forecasting methods. International Journal of Forecasting. 2018;**36**(1):54-74

[24] Taylor JW. Rethinking the evaluation of model uncertainty via weighted likelihoods. Journal of Business & Economic Statistics. 2020;**38**(2): 386-394

[25] Zheng Y, Lee HS, Jordan S. Support vector regression for short-term solar power forecasting. Solar Energy. 2013; **92**:274-283

[26] Inman RH, Pedro HT. Solar Energy Forecasting and Resource Assessment. United States: Academic Press; 2018

Chapter 3

Parameter Identification of Solar Cell Mathematical Models Using Metaheuristic Algorithms

Hamdy M. Sultan, Mahmoud A. Mossa and Almoataz Y. Abdelaziz

Abstract

Lately, there has been a notable interest among researchers in constructing a precise mathematical representation using experimentally gathered data from solar cells and photovoltaic (PV) modules. This representation serves as a means to simulate and assess the performance of PV systems. In this study, the Walrus Optimization Algorithm (WaOA) and Cheetah optimizer (CO) were employed to deduce the unknown parameters inherent in various modes of solar cells and PV modules, specifically the single-diode model (SDM) and double-diode model (DDM). Furthermore, the evaluation criterion for this work involved measuring the route mean square error (RMSE) between the simulated outcomes generated using identified parameters for each mathematical model and the actual voltage derived from measurements of solar cells and PV modules. Notably, a comprehensive statistical analysis was carried out to validate the efficacy and stability of the WaOA and CO algorithms. These algorithms were compared against other optimization techniques for their effectiveness in solving the optimization challenge of accurately estimating the design parameters of PV systems. The outcomes of simulations and the extensive statistical assessment substantiate the superior performance and reliability of the Walrus Optimization Algorithm in effectively extracting parameter values from diverse PV modules under various operational scenarios.

Keywords: optimization, solar cells, PV modules, walrus optimization algorithm, cheetah optimizer, SDM, DDM

1. Introduction

The finite nature of fossil fuel reserves and their excessive utilization have not only put human health at risk but have also had detrimental effects on the ecological balance. Hence, the imperative to develop sustainable energy alternatives has become exceedingly urgent [1, 2]. Among the array of renewable energy options available such as solar, hydro, wind, geothermal, and biomass energy, these sources are either practically inexhaustible or can be replenished in the short term [1, 3].

IntechOpen

Solar energy, in particular, stands as a formidable candidate, possessing an immense reservoir of energy capable of satisfying the entire spectrum of contemporary human energy needs. This potential has propelled the integration of solar energy into diverse applications, including desalination processes, heating installations, and the generation of photovoltaic (PV) power [4–6]. Given its dual attributes of cleanliness and widespread accessibility, PV power generation has evolved into a pivotal initiative within the larger framework of advancing renewable energy sources [3].

Notably, the pursuit of PV power generation aligns seamlessly with the growing demand for clean electrical energy across various sectors. As a result, it has assumed a critical role in the ongoing efforts to cultivate and expand renewable energy alternatives. The trajectory of shifting toward PV power generation underscores a proactive stance in mitigating the environmental repercussions of fossil fuel reliance while fostering a sustainable energy landscape for the future [7, 8].

Doping, the intentional introduction of impurities into semiconductor materials, plays a pivotal role in modulating electronic properties, and understanding its effects is essential for precise solar cell parameter extraction. The study presented in [9] explores the correlation between secondary electron doping contrast and Fermi level pinned surfaces in silicon samples, prepared through HF-based wet-chemical treatment or focused ion beam micromachining. This investigation employs energy-resolved SE imaging techniques and finite element analyses, revealing surface band-bending effects and challenging the conventional belief exclusively associating doping contrast with patch fields or adventitious metal-semiconductor contacts. In a complementary approach, Chee et al. [10] introduce a Monte Carlo model incorporating a finite element method and a ray tracing algorithm to detail the computation of electrostatic fields inside and outside a semiconductor for doping contrast in a scanning electron microscope (SEM). This numerical method effectively distinguishes effects on doping contrast arising from surface band bending, external patch fields, and macroscopic external fields in the SEM detection system. The presented theory aligns well with experimental observations, advancing our understanding of doping contrast mechanisms and facilitating quantitative dopant profiling using the SEM. Building on this foundation, Chee [11] discusses doping contrast utilization with the secondary electron (SE) signal in the SEM for quantitative dopant profiling in alignment with International Roadmap for Semiconductors (ITRS) requisites. This research specifically focuses on site-specific dopant profiling of silicon p–n junction specimens, employing a 30-kV Ga^+ focused ion beam (FIB) for trench side-wall cutting and successive milling at lower voltages in a dual-beam FIB/SEM system. Despite the protective platinum strap's effective control of "curtaining" 'effects, reduced doping contrast from the side wall, compared to a cleaved surface with the same ion beam energy, is attributed to material effects from prior milling steps and differences in geometries between milling and imaging. The study introduces novel principles underlying the doping contrast mechanism, considering ion implantation depth and concentration, and amorphization damage as linear functions of the final milling voltage. Although reaching only half the contrast achievable from a freshly-cleaved specimen, the research demonstrates the feasibility of site-specific dopant profiling *in situ* in the SEM with doping contrast increasing as milling voltage decreases. The continual advancement of semiconductor fabrication processes stands as a cornerstone in the relentless pursuit of highly efficient inorganic photovoltaic (PV) technologies. Through dedicated research and development efforts, these processes have undergone refinement, unlocking new frontiers in the realm of solar energy harnessing. This evolution is pivotal in enhancing the overall performance, durability, and cost-effectiveness of inorganic PV systems.

The effective design, forecasting, sizing, diagnostics, and maintenance of photovoltaic (PV) system installations rely heavily on a dependable and precise model encompassing the cell, module, and PV array components [12, 13]. In the realm of literature, three distinct models have been established, each featuring one, two, or three diodes. These models are characterized by parameters that demand accurate extraction. The process of obtaining these diverse parameters remains both critical and complex. To tackle this challenge, numerous techniques have emerged in scholarly works to deduce the optimal PV parameters. These techniques span numerical, analytical, evolutionary, and hybrid methodologies [14, 15].

Among the various methods available for parameter extraction in PV systems, the equation governing the current-voltage characteristic holds particular promise. This equation encapsulates the entirety of PV properties and parameters. However, its transcendental nature introduces complexities in solving it, transforming it into an optimization predicament. In navigating this complexity, metaheuristic approaches have surfaced as particularly effective tools. Their demonstrated success in addressing a spectrum of challenges across diverse domains attests to their potential in resolving the intricacies of PV parameter optimization. As such, these metaheuristics offer a promising avenue to surmount the complexities inherent in solving the multifaceted equation governing PV behavior, advancing the domain of PV system analysis and application [16, 17].

The precision of modeling holds significant importance in the evaluation, enhancement of efficiency, fault analysis, and simulation of photovoltaic (PV) systems. These systems, composed of aggregated PV cells, are typically subject to modeling through equivalent circuits, encompassing single-diode (SDM), double-diode (DDM), and triple-diode (TDM) models [18]. These circuit representations effectively emulate the electrical attributes of PV cells. In these models, parameters are present in quantities of five, seven, and nine, respectively, demanding precise extraction [19].

Intricacies arise with the expansion of diode numbers, which introduces a greater number of parameters needing extraction, thus intensifying computational complexities. The challenges posed encompass not only the heightened intricacy of solutions due to multiple unknown variables but also the interdependence between electrical values, which results in a highly implicit function. Furthermore, the incorporation of exponential functions within the characteristic equations exacerbates the difficulty in solving nonlinear attributes [20, 21].

These compounded challenges coalesce to create a perplexing puzzle when it comes to establishing accurate PV models. This enigma requires innovative approaches that can surmount the intertwined complexities of parameter extraction and nonlinear behavior to arrive at models that truly reflect the intricacies of PV systems [21].

In the research conducted by Senturk and Eke [22], a novel empirical relationship was utilized to extract parameters within the framework of the one-diode model. This empirical approach involves determining the initial value of the series resistance using the slope of the I-V curve as given by the producer. However, acquiring numerical data for the current-voltage characteristic is often not readily available upon purchasing a photovoltaic (PV) module. To address this limitation, image processing techniques are necessitated to extract the data from technical documentation, yet the accuracy of this extracted data may vary.

Analytical methods exhibit effectiveness under standard weather conditions; however, their reliability diminishes in the face of fluctuating atmospheric conditions

[23]. Furthermore, approximating equations in these methods substantially compromises result accuracy. On a different note, Tossa et al. [24] introduced an innovative approach for accurately modeling the SD model of a PV module. This method, implemented in the MATLAB/Simulink environment, relies on the Levenberg-Marquardt algorithm. Another avenue, proposed by Ghani et al. [25], involves scrutinizing the current-voltage characteristic to derive parameter values. Here, a model of five formulas is solved utilizing the Newton-Raphson method. Yet, this algorithm not only requires solving a system of equations but also entails handling a Jacobian matrix composed of twenty-five first and second derivative terms, adding a layer of complexity to the process.

The notable drawback of numerical methods, such as the Newton-Raphson, is their demand for extensive calculations to ensure convergence. This method's effectiveness diminishes as the number of parameters that have to be identified rise. Despite their efficiency, the sluggish convergence of numerical techniques does not always guarantee optimal outcomes as they might converge to a local minima. Additionally, selecting appropriate initial conditions for these methods can be challenging [23]. In light of these complexities, the field continues to seek innovative solutions that can offer improved accuracy and efficiency in parameter extraction while addressing the limitations of current techniques.

In the study by Oliva et al. [25], a novel approach utilizing a chaotic artificial bee colony (ABC) algorithm was introduced to estimate parameters for photovoltaic (PV) panels. The innovation in this proposed chaotic ABC lies in the replacement of the arbitrary number in the onlooker bee phase with a number produced through chaos theory. This chaotic number is obtained from a carefully selected chaotic map, with the tent map being identified as the optimal choice and subsequently incorporated into the proposed chaotic ABC algorithm. A comprehensive exploration of various chaotic maps was conducted, culminating in the selection of the tent map. Yang introduced a novel, versatile population-based optimization technique called hunger games search (HGS) in 2021 [26]. This optimizer demonstrated scalability and adaptability, making it suitable for a wide range of optimization challenges spanning application and structural domains. A notable feature of HGS was its Laplacian-based crossover search mechanism, which effectively diversified solution exploration. Additionally, the integration of the Nelder-Mead (NM) local search technique heightened the precision of convergence toward the optimal solution. Building upon the original HGS framework, the study extended its capabilities by incorporating the Laplacian mechanism and the Nelder-Mead simplex strategy. This extension resulted in an enhanced parameter optimization performance, successfully applied to the realm of static photovoltaic models. The proposed method emulated the hunger-induced behavior and attributes of individuals within the optimization process. Despite HGS finding utility in various engineering problems, its potential in parameter estimation for solar PV models had remained untapped. To address this gap, a new algorithm emerged, combining elements from the simplified swarm optimizer (SSO) with the Nelder-Mead simplex approach. This innovative fusion led to a dual achievement of precise and swift parameter identification for both static deflection model (SDM) and dynamic deflection model (DDM). The novel approach exhibited the capability to finely balance accuracy and speed in parameter estimation, demonstrating its potential for practical applications in the solar PV domain [26].

The simulation outcomes, utilizing both SD and DD models, clearly underscore the advantages of the developed chaotic ABC over a range of competing techniques. These include the conventional ABC, a chaotic variant of the particle swarm optimization

(PSO), the artificial bee swarm optimization algorithm, simulated annealing, the cat swarm optimization algorithm, teaching learning-based optimization algorithm, and the harmony search method. To provide concrete context, the study employed real-world case studies involving an RTC-France PV cell, a polycrystalline panel labeled STM6-120/36, and a monocrystalline PV panel denoted as STM6-40/36.

The utilization of the chaotic ABC technique presents a significant step forward in enhancing the accuracy and efficiency of parameter estimation for PV panels. By harnessing the dynamics of chaos theory, this approach demonstrates its capacity to outperform several established optimization methods. As the realm of PV technology continues to evolve, these innovative methodologies hold the potential to substantially contribute to the improvement of PV system performance, thereby promoting the broader adoption of renewable energy sources.

The work by Oliva et al. [27] presents a notable advancement by introducing a chaotic variation of the whale optimization algorithm (WOA) for the precise estimation of parameters associated with photovoltaic (PV) cells and panels. The foundation of the WOA draws inspiration from the behavioral patterns of whales. In the WOA framework, during each iteration, the repositioning of whales involves the generation of a random number within the range of [0,1]. This number then dictates the likelihood of selecting either a spiral model or a shrinking encircling mechanism for adjustment. The innovation in the proposed chaotic WOA lies in the replacement of this randomly generated number with a value generated through chaos theory.

In the pursuit of the optimal chaotic map for this endeavor, four distinct chaotic maps—namely, Singer, Sinusoidal, Logistic, and Tent—underwent testing. Through rigorous evaluation, the singer map emerged as the most fitting choice for integration into the chaos-based WOA, proving its mettle in tackling the parameter estimation challenge effectively. Simulation outcomes, encompassing both single- and double-diode models, serve as compelling evidence of the superiority of the proposed chaos-infused WOA over a spectrum of cutting-edge optimization algorithms. Among these algorithms are the ABC, PSO, artificial bee swarm optimization, simulated annealing, bird mating optimization, differential evolution, and harmony search. These results underscore the potential and advantages of this chaotic WOA variant, showcasing its ability to surpass established optimization techniques in addressing the complexities of parameter estimation in PV systems. By harnessing the power of chaos theory within optimization frameworks, this research contributes to the refinement of parameter estimation processes within the domain of photovoltaics. Such innovative methodologies are integral to advancing the efficiency and effectiveness of PV systems, aligning with the broader global push for sustainable and renewable energy sources.

The study conducted by Mughal et al. [28] introduces an innovative approach that merges the strengths of PSO and SA to estimate parameters related to PV cells. In the proposed PSO-SA hybrid framework, the advantages of both algorithms are combined by applying the operators of PSO and SA to the search agents during each iteration. The optimization algorithm employs the RMSE as its primary objective. The simulation outcomes, which involve an RTC-France PV cell encompassing both single- and double-diode models, unequivocally demonstrate the superior performance of the proposed hybrid algorithm when compared to PSO, SA, harmony search (HS), and pattern search techniques.

Furthermore, in Gong et al. [29], a distinct application of PSO is presented. This version utilizes varying acceleration coefficients and inertia weights for parameter extraction of PV modules. In the research by Fathy and Rezk [30], an imperialistic competitive algorithm (ICA) is leveraged to estimate circuit model parameters associated with PV modules. The investigation spans both single- and double-diode models, considering various PV technologies such as mono-crystalline, polycrystalline, thin-film, and amorphous materials. The incorporation of amorphous on crystalline materials into the analysis signifies recognition of the evolving landscape of PV technologies. This nuanced exploration becomes particularly pertinent in instances where focused ion beam (FIB) processing of crystalline materials is employed, resulting in hybrid structures that merge the characteristics of amorphous and crystalline semiconductors. The interplay between these materials introduces a unique set of challenges and opportunities, making them a crucial consideration in the development and application of single- and double-diode models [31]. The attainments underscore the superiority of ICA over several alternative optimization techniques, including pattern search, chaos-based PSO, bird mating optimization, adaptive differential evolution (DE), ABSO, SA, and HS [30].

These studies collectively highlight the ongoing quest for refining parameter estimation methodologies in the realm of photovoltaics. The fusion of optimization algorithms and the exploration of their variations underscore the significance of devising novel strategies to enhance the accuracy and efficiency of PV system performance analysis. As the renewable energy landscape continues to expand, the advancements in parameter estimation techniques contribute to the broader mission of harnessing sustainable energy sources and minimizing environmental impact. In addition to the research mentioned, other studies have also employed distinct algorithms for the parameter extraction of photovoltaic cells. For instance, Ram et al. [32] introduced an enhanced version of the flower pollination algorithm, reference [33] utilized the fireworks optimization technique, and Abdelghany et al. [34] employed the water cycle algorithm for this purpose. While metaheuristics offer several advantages over alternative parameter estimation methods, it is important to acknowledge a potential pitfall known as premature convergence, especially when dealing with complex PV cell parameter estimation issues, which often exhibit multimodal characteristics.

Premature convergence occurs when metaheuristic algorithms become trapped in local optima instead of reaching the global optimum. This concern is particularly relevant in the context of PV cell parameter estimation, where the problem's multi-peaked nature can hinder the ability of algorithms to find the most accurate parameter values. This challenge is further evidenced by the relatively high root mean square error (RMSE) values observed in the application of various metaheuristic algorithms to PV cell parameter estimation [35].

To address this issue of premature convergence, extensive efforts are being dedicated to developing strategies that enhance the performance of metaheuristic techniques in the context of PV cell parameter estimation. Researchers are actively exploring techniques to ensure that these algorithms have a higher likelihood of escaping local optima and converging toward more accurate parameter values. By mitigating the premature convergence problem, the potential of MAs can be maximized in accurately estimating the parameters of PV cells and modules.

In this chapter, the investigation revolves around the utilization of both the Walrus Optimization Algorithm (WaOA) [36, 37] and the Cheetah optimizer (CO) [38] as potent tools for addressing the intricate challenge of PV cell parameter estimation. By employing these advanced optimization techniques, the aim is to enhance the accuracy and efficiency of estimating parameters crucial to the performance of photovoltaic cells. The Walrus Optimization Algorithm, characterized by its unique nature-inspired strategies, and the Cheetah optimizer, drawing inspiration from the swift and agile cheetah's hunting behaviors, are harnessed as innovative approaches to tackle the complexities inherent in this parameter estimation problem.

The remainder of this chapter follows a structured organization. In Section II, the formulation of the PV cell parameter estimation problem is presented. Section III introduces the methodologies that have been proposed. Moving forward, the fourth section presents the outcomes and analysis. Ultimately, the fifth section contains the concluding remarks to wrap up the discussion.

2. Models of solar cells

A mathematical model must be utilized to determine the solar cell properties analytically in order to construct the solar cells and PV modules. Diode-based electronic circuits are employed to model solar cells based on this. Along with the TD circuit, which was very recently developed, the SD and DD models are the most often used tools for determining the parameters of solar cells and PV modules [38].

2.1 Single-diode model

The model, which can be seen in **Figure 1**, simply has one diode that is utilized to parallelize the current source I_{ph}, which displays the photo-generated electrical current. The diode functions as a rectifier of half-waves. The nonphysical accessibility factor of the ideality of the diode is taken into account by the mathematical framework [39]. The model has a fairly straightforward structure, making it simple to put into practice. The biggest problem with this straightforward model is that there are only five parameters that need to be carefully specified.

According to **Figure 1**, I_t can be expressed as;

$$I_t = I_{ph} - I_{sd} - I_{sh} \tag{1}$$

where I_t, I_{ph}, I_{sd}, and I_{sh} imply the total current, photo current, diode current, and current through shunt resistance, respectively.

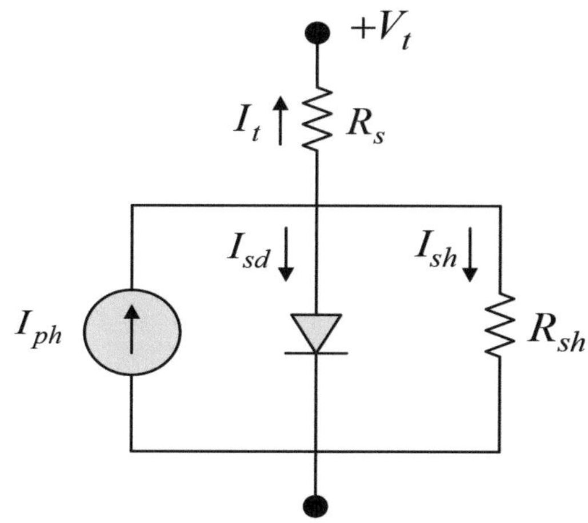

Figure 1.
Equivalent circuit for SD model.

In order to create a more accurate model for the inside functioning of the diode, Shockley's equivalent diode formula might be used. Eq. (1) is capable of being rewritten as follows [39, 40]:

$$I_t = I_{ph} - I_{sd}\left[exp\left(\frac{q(V_t + R_s I_t)}{nkt}\right) - 1\right] - \frac{V_t + R_s I_t}{R_{sh}} \tag{2}$$

Where V_t implies the model terminal voltage, I_{sd} implies reverse saturation current of the diode, R_s implies the resistance in series connection, R_{sh} implies the resistance in the parallel combination, n denotes the ideality factor, q implies the electron charge ($q = 1.602 * 10^{-23}$ J/K), k implies the Boltzmann constant ($k = 1.380 * 10^{-23}$ J/K), and T implies cell temperature i. Therefore, precise estimation of these variables, which will be carried out using various optimization strategies in the subsequent sections, is necessary to ensure the model operates properly.

2.2 Double-diode model

The double-diode (DD) model, shown in **Figure 2**, is suggested as a replacement for the single-diode model, which is typically not a good option for use with a variety of applications [41]. The solar spectrum containing sufficient energy irradiating the p-n junction can induce a photocurrent that changes the junction potential. The excess electron-hole pairs generated can forward bias the p-n junction [42, 43]. The surface or interface traps/charges also play a crucial role in the nonideal behavior of the solar cell [44, 45], as well as the carrier distributions following the electric potentials influenced by the solar irradiation [46].

As seen in **Figure 2**, there are two diodes: one serves as a rectifier, and the other is utilized to account for the effects of the solar cell's non-idealities plus the current generated from the combination. The following equation could be used to depict the current balance in the corresponding circuit of **Figure 2** [40]:

$$I_t = I_{ph} - I_{d1} - I_{d2} - I_{sh} \tag{3}$$

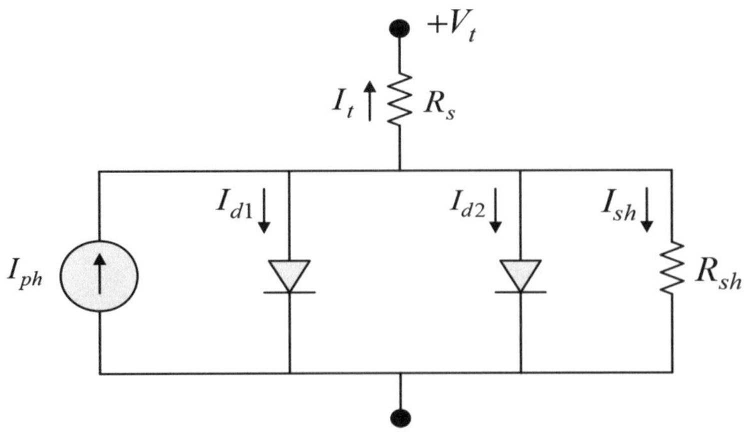

Figure 2.
Equivalent circuit for DD model.

where I_{d1}, and I_{d2} signify current of the first and second diodes. The two diodes' internal configuration is updated using the Shockley equivalency. Therefore, Eq. (3) can be rewritten as given [40],

$$I_t = I_{ph} - I_{d1}\left[exp\left(\frac{q(V_t + R_s I_t)}{n_1 kt}\right) - 1\right] - I_{d2}\left[exp\left(\frac{q(V_t + R_s I_t)}{n_2 kt}\right) - 1\right] - \frac{V_t + R_s I_t}{R_{sh}} \tag{4}$$

where I_{d1} and I_{d2} symbolize saturation currents for the two diodes in the circuit. n_1, n_2 signify ideality factors. The DD model contains seven unidentified parameters that have to be precisely determined, that is, R_s,R_{sh}, I_{ph}, I_{d1}, I_{d2},n_1, and n_2.

2.3 Problem formulation

It is feasible to approach each of the mathematical models for the SD and DD as an optimization issue, with the best values for the unidentified model parameters serving as the answer. To determine if the optimized parameters meet their real values or not, a fitness function needs to be initially applied. Examining the level of agreement between the I-V curves listed in the datasheet of an actual solar cell and those anticipated regarding the identified parameters of the theoretical empirical model is another way to determine the quality of the methodology used for estimation. Consequently, the fitness function for the SD can be illustrated as [37, 41]:

$$f_{SD}(V_t, I_t, x) = I_t - x_3 + x_4\left[exp\left(\frac{q(V_t + x_1 I_t)}{x_5 kt}\right) - 1\right] - \frac{V_t + x_1 I_t}{x_2} \tag{5}$$

meanwhile, for the DD model, the fitness function is written as:

$$f_{DD}(V_t, I_t, x) = I_t - x_3 + x_4\left[exp\left(\frac{q(V_t + x_1 I_t)}{x_6 kt}\right) - 1\right] + x_5\left[exp\left(\frac{q(V_t + x_1 I_t)}{x_7 kt}\right) - 1\right] - \frac{V_t + x_1 I_t}{x_2} \tag{6}$$

Where $V_t, and\ I_t$ imply the experimental values of voltage and current of the solar cell. For SD model, $x = [R_s, R_{sh}, I_{ph}, I_{sd}, n]$ symbolizes the solution vector, wherein the case of the DD model the solution vector is $x = [R_s, R_{sh}, I_{ph}, I_{d1}, I_{d2}, n_1, n_2]$. The functions f_{SD}, f_{DD} assess and investigate how closely each circuit's outcomes resemble those obtained through experimental measurement.

The cost function is then developed based on identifying the parameters that provide the least amount of error between the actual, as determined by observations, and the predicted, as determined by diode models. The cost function can then be defined using a set of NE samples to broaden the search until global optima are found and expressed as:

$$RMSE(x) = \sqrt{\frac{1}{N}\sum\nolimits_{c=1}^{NE}\left(f_M^c\left(V_t^c, I_t^c, x\right)\right)^2} \tag{7}$$

where M aids in determining, which diode model should be used and RMSE stands for root mean square error. The Walrus Optimization Algorithm (WaOA) and

Cheetah optimizer (CO) are employed to determine the best values for the design parameters of various PV modules and solar cells that produce the objective function's least value.

3. Optimization algorithms

3.1 Walrus optimization algorithm (WaOA)

A large marine mammal with flippers, the walrus has a patchy distribution in the Arctic Ocean and the waters surrounding the North Pole in the Northern Hemisphere [35, 36]. Large tusks and whiskers on adults make them easy to recognize. Walruses are sociable animals who spend the most of their time searching for benthic bivalve mollusks to feed on the sea ice. The tusks on walruses, which are this animal's most noticeable trait, are very long. These are elongated canines that are present in both male and female animals and can grow to be up to 1 m long and 5.4 kg in weight. In the late summer, walruses prefer to move to outcrops or rocky beaches when the weather warms and the ice begins to melt. These migrations are quite dramatic and include large gatherings. Due to their size and tusks, the polar bear and the killer whale (orca) are the walrus' only two natural predators.

Walruses exhibit intellectual behavior in their social interactions and everyday actions. Three of these clever actions stand out as being the most inescapable:

1. *Directing the individuals to eat under the direction of the tribe member with the longest tusks:* During the search process, the algorithm is guided toward areas of greatest potential by keeping track of the best population participant. The dominant walrus, which is distinguished by having a longer tusk, is in charge of leading the other walruses in the social life of the species. This method involves moving walruses, which significantly alters their posture. The algorithm's capacity for global search and exploration is improved by simulating these huge displacements [35, 36].

2. *Migration toward rocky beaches:* The migration of walruses in response to summer's warming temperatures is one of their normal behaviors. Walruses shift dramatically in this process, going toward rocky outcrops or beaches. The position of other walruses is taken into account as a walrus's migration destination in the WaOA simulation for a walrus. One of these locations is arbitrary pointed out, and the walrus migrates in its direction [35, 36]. WaOA's architecture, which follows this tactic, enhances the possibilities of global search and discovery. The foraging approach used by the strongest walrus differs from the migratory strategy in that the population update process is not allowed to rely just on one individual, such as the population's best member. This updating procedure stops the algorithm from stalling in local optima as well as premature convergence.

3. *Fight or run away from predators:* When battling their predators, such as polar bears and killer whales, walruses adopt a protracted chasing technique. The walrus position is only little altered by this chasing activity, which occurs in a tiny area nearby. WaOA's capacity to seek locally and exploit to converge to better solutions is therefore improved by imitating the minor movements of the walrus by looking for good locations throughout the struggle.

3.1.1 Initialization

Walruses serve as the search population in WaOA, and a potential answer to the optimization problem is represented by each walrus in WaOA. The potential values for the problem variables are thus determined by each walrus' location inside the search space. Walrus populations are arbitrarily initialized at the start of WaOA deployment. Utilizing (8), we arrive at this WaOA population matrix [35, 36].

$$X = \begin{bmatrix} X_1 \\ \vdots \\ X_i \\ \vdots \\ X_N \end{bmatrix} = \begin{bmatrix} x_{1,1} & \cdots & x_{1,j} & \cdots & x_{1,m} \\ \vdots & \ddots & \vdots & ⋰ & \vdots \\ x_{i,1} & \cdots & x_{i,j} & \cdots & x_{i,m} \\ \vdots & \ddots & \vdots & ⋰ & \cdots \\ x_{N,1} & \cdots & x_{N,j} & \cdots & x_{N,m} \end{bmatrix} \tag{8}$$

Where N denotes the number of walruses, m denotes the number of choice variables, X signifies the population of walruses, X_i signifies the ith walrus (candidate solution), and $x_{i,j}$ signifies the value of the jth decision variable recommended by the ith walrus. For the fitness function derived from walruses, the generated values are described in (9).

$$F = \begin{bmatrix} F_1 \\ \vdots \\ F_i \\ \vdots \\ F_N \end{bmatrix} = \begin{bmatrix} F(X_1) \\ \vdots \\ F(X_i) \\ \vdots \\ F(X_N) \end{bmatrix} \tag{9}$$

where F signifies the vector of the fitness function and F_i signifies the value of the fitness function calculated depending on the ith walrus.

3.1.2 Mathematical modeling of WaOA

3.1.2.1 Phase 1: feeding strategy (exploration)

Walruses eat a wide variety of marine creatures, including more than 60 different species of sea cucumbers, tunicates, soft corals, tube worms, prawns, and different mollusks. The walrus prefers benthic bivalve mollusks, particularly clams, and forages by grazing on the ocean floor while using its active flipper movements and sensitive vibrissae to search out and find food. The more powerful walrus in the group, the one with the largest tusks, leads the other walruses in their search for nourishment [35, 36]. The degree of accuracy of the fitness values of the potential solutions is comparable to the size of the tusks in walruses. The most powerful walrus in the team is pointed out as the best candidate solution having the best fitness. The WaOA's exploration effectiveness in the global search is increased as a result of the walruses' searching behavior, which results in varied scanning regions of the search field. Following the direction of the most important member of the group, (10) and (11) are used to mimic the process of changing the walruses' position depending on the grazing strategy [35, 36]. A new position for the walrus is first created in this process using (10). If this novel location increases the fitness value, it replaces the earlier location; this idea is modeled in Eq. (11).

$$x_{i,j}^{P_1} = x_{i,j} + rand_{i,j} \cdot (SW_j - I_{i,j} \cdot x_{i,j}) \tag{10}$$

$$X_i = \begin{cases} X_i^{P_1}, & F_i^{P_1} < F_i \\ X_i, & else \end{cases} \tag{11}$$

where $X_i^{P_1}$ denotes the updated position for the ith walrus according to the feeding stage, $x_{i,j}^{P_1}$ denotes its jth dimension, $F_i^{P_1}$ denotes the fitness value, $rand_{i,j}$ signifies arbitrary numbers ranged from 0 to 1, SW denotes the finest potential solution related to the most powerful walrus, and $I_{i,j}$ is an integer number selected randomly between 1 and 2. $I_{i,j}$ is utilized to enhance the algorithm's exploration capability so that if it is selected as 2, it creates more significant and broader changes in the location of candidates compared to the case when the value is selected as 1, which presents the normal condition of this displacement. These conditions help enhance the global searching of the WaOA in escaping from the local optima and finding the global optimal spot in the problem-solving area.

3.1.2.2 Phase 2: migration

Due to the rising temperature of the air in the final days of summer, one of the normal behaviors of walruses is their movement to outcrops or rocky beaches. The WaOA uses this migratory process to direct the walruses in the search space to find appropriate regions in the search space [32, 33]. Employing (12) and (13), this behavioral process is quantitatively modeled. This modeling makes the assumption that each walrus moves to a different walrus location (chosen at random) in a different region of the search field. As a result, the suggested alternative location is initially derived using (12). Then, in accordance with (13), the most recent position supersedes the earlier one of the walrus if it increases the fitness value [35, 36].

$$x_{i,j}^{P_2} = \begin{cases} x_{i,j} + rand_{i,j} \cdot (x_{k,j} - I_{i,j} \cdot x_{i,j}), & F_k < F_i \\ x_{i,j} + rand_{i,j} \cdot (x_{i,j} - x_{k,j}), & else \end{cases} \tag{12}$$

$$X_i = \begin{cases} X_i^{P_2}, & F_i^{P_2} < F_i \\ X_i, & else \end{cases} \tag{13}$$

where $X_i^{P_2}$ denotes the updated location of the ith walrus according to the migration stage, $x_{i,j}^{P_2}$ denotes the jth dimension, $F_i^{P_2}$ denotes the fitness value, X_k, $k \in \{1, 2, \ldots, N\}$ and $k \neq i$, signifies the position of the pointed walrus toward the ith walrus will move, $x_{k,j}$ denotes the jth dimension, and F_k denotes its fitness value.

3.1.2.3 Phase 3: running out and fighting against predators (exploitation)

Polar bear and killer whale assaults are constant threats to walruses. The walruses move around in the area around where they are located as a result of their technique of evading and combating these predators. The WaOA's ability to utilize this aspect of walrus behavior in the small search area surrounding potential solutions is enhanced. Since this process takes place close to each walrus's location, in the WaOA algorithm, it is considered that this range of walrus location change takes place in a neighborhood centered on walruses that has a specific radius. The radius of this neighborhood is thought of as a variable because it initially begins at its greatest value and then decreases throughout the algorithm repetitions because, in the early iterations, priority is given to global search to attempt to find the ideal area in the search field. Because of this, a variable radius with algorithm iterations has been created in this stage of WaOA using localized lower/upper boundaries [34, 35]. For the purpose of simulating the scenario in WaOA, a neighborhood is assumed to be present around each individual walrus, which is then given an arbitrary new location inside that neighborhood using steps (14) and (15). Then, in accordance with (16), this new location substitutes the prior one if the fitness value is enhanced [35, 36].

$$x_{i,j}^{P_3} = x_{i,j} + \left(lb_{local,j}^{t} + \left(ub_{local,j}^{t} - rand \cdot lb_{local,j}^{t}\right)\right), \tag{14}$$

$$Local\ Bounds : \begin{cases} lb_{local,j}^{t} = \dfrac{lb_j}{t}, \\ ub_{local,j}^{t} = \dfrac{ub_j}{t}, \end{cases} \tag{15}$$

$$X_i = \begin{cases} X_i^{P_3}, & F_i^{P_3} < F_i \\ X_i, & else \end{cases} \tag{16}$$

where lb_j and ub_j denote the lower and upper boundaries of the jth variable, $lb_{local,j}^{t}$and $ub_{local,j}^{t}$ denote local lower and local upper boundaries allowable for the jth variable. $X_i^{P_3}$ denotes its jth dimension. $F_i^{P_3}$ denotes its fitness value. **Figure 3** displays the flowchart for implementing the WaOA.

3.2 Cheetah optimizer (CO)

3.2.1 Inspiration

The cheetah (*Acinonyx jubatus*) stands out as the primary feline species and is recognized as the fastest land animal, inhabiting the central regions of Iran and Africa. These remarkable creatures are capable of reaching speeds exceeding 120 kilometers per hour. Their exceptional speed and agility are attributed to their physical characteristics, including a long tail, slender legs, lightweight build, and a flexible spine. Cheetahs are known for their swiftness, adept stealth, rapid pursuit during hunting, and distinctive spotted coats. However, it is important to note that these visually-oriented predators cannot sustain their high-speed actions for extended periods, limiting their chases to less than half a minute [37]. After capturing their prey, the cheetah's speed dramatically drops from 93 kilometers per hour (58 mph) to 23 kilometers per hour (14 mph) in just three strides. To overcome this limitation, cheetahs employ a keen sense of observation, often perching on small branches or hills to scan their surroundings for potential prey. Their ability to blend seamlessly into high, dry grass due to their unique coat pattern further aids in successful hunting. Cheetahs typically target animals such as Thomson's gazelles, impalas, antelopes, hares, birds, rodents, and young calves of larger herbivores [37]. The hunting strategy

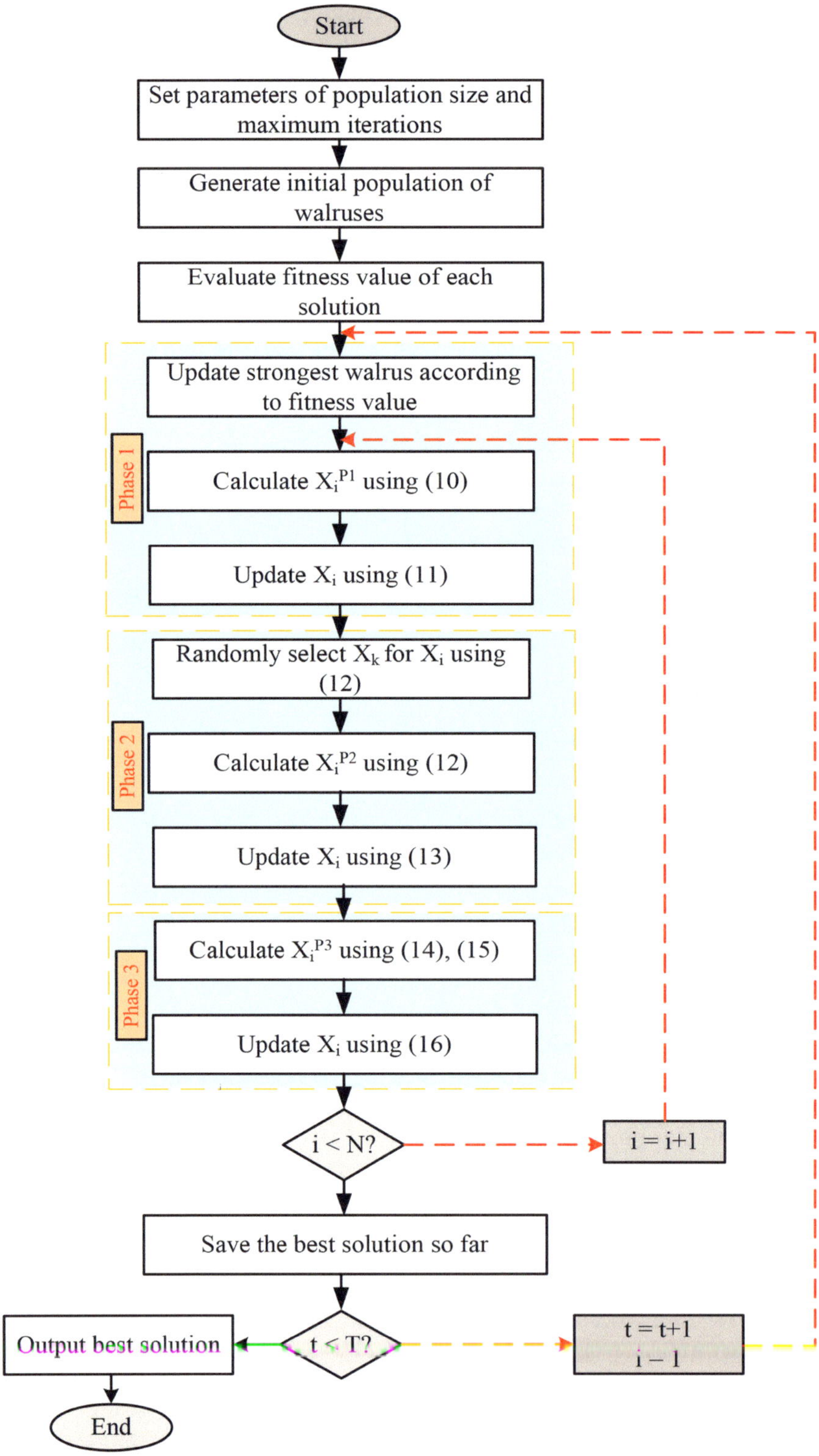

Figure 3.
Flowchart of WaOA technique.

of cheetahs involves slow, stealthy approaches to minimize the distance between themselves and their prey. They maintain a crouched posture and patiently wait for the prey to draw near as they tend to abandon the hunt if detected. The preferred minimum distance for a successful chase ranges from 60 to 70 m (200 to 230 feet), but it increases to 200 meters (660 feet) if they fail to remain concealed adequately. The pursuit itself typically lasts about 60 seconds, covering an average distance ranging from 173 m (568 feet) to 559 m (1834 feet).

To bring down their prey, cheetahs employ a tactic involving a swift forepaw strike to the prey's rump, causing the prey to lose its balance. Subsequently, the cheetah utilizes its strength to overpower the prey and affect a swift kill. The cheetah's muscular tail plays a pivotal role in achieving sharp turns during the pursuit. Generally, it is easier for them to hunt animals that have strayed from their herds or display less vigilance. Predation outcomes are influenced by various factors such as the age and gender of the cheetah, the number of predators involved, and the level of alertness displayed by the prey. Additionally, coalitions of cheetahs or mothers with cubs tend to be more successful in hunting larger prey.

Biological studies have revealed that cheetahs possess extraordinary spinal flexibility and long tails that contribute to their physical balance. Their shoulder blades, which are not connected by collarbones, enable a wider range of shoulder movement, enhancing their hunting prowess. Despite these exceptional attributes, it is important to acknowledge that not all cheetah predations result in a successful capture.

3.2.2 Mathematical model and algorithm

It is probable to spot prey when a cheetah patrols or examines its immediate surroundings. The cheetah may sit in one spot after spotting its victim, watch until it approaches, and then launch an assault. There are phases of rushing and capturing in the attack phase. The cheetah might cease hunting for a number of reasons, including power limitations, the ability to catch prey quickly, etc. The CO algorithm's overall foundation is the intelligent application of different hunting techniques during hunting sessions [37].

- *Searching:* Cheetahs must seek, either actively or passively, throughout their territory (search space) or the surrounding region in order to find their prey.

- *Sitting and waiting:* If the prey is discovered but the circumstances are not ideal, cheetahs may wait for the prey to approach or for the situation to improve;

- *Attacking:* This tactic entails two crucial steps.

    - *Rushing:* The cheetah will move as quickly as possible toward its victim when it intends to attack.

    - *Capturing:* The cheetah approached its prey while moving quickly and maneuverably.

- *Leave the prey and return home:* Two scenarios are taken into consideration for this tactic. (1) If the cheetah is unable to catch its prey, it should move or go back to

(a) (b) (c) (d)

Figure 4.
Hunting behaviors of cheetah: (a) Scanning, (b) sitting and waiting, (c) rushing, and (d) capturing.

its home zone. (2) It may move to the location of the most recent prey found and search the area around it if there has been no successful hunting activity for a period of time (**Figure 4**).

3.2.3 Searching strategy

Cheetahs hunt for prey using one of two methods: either vigorously patrolling the territory while sitting or standing to scan the surroundings. When the prey is numerous and grazing while traversing the plains, scanning manner is more appropriate. However, if the prey is dispersed and active, it is preferable to choose an active style that consumes much power than the scan method. A series of such searching strategies may, therefore, be chosen by the cheetah throughout the time of hunting, subject to the state of the prey, the area's coverage, and the cheetahs' own health [37].

Mathematically, the cheetah's states (other configurations) create a population, and each victim is a spot of a decision variable matching to the best option (see **Figure 5a**). Then, using an arbitrary step size and the present position of each cheetah in the arrangement as a starting point, this equation is suggested:

$$X_{i,j}^{t+1} = X_{i,j}^{t} + r_{i,j}^{-1} \cdot \alpha_{i,j}^{t} \tag{17}$$

Where $X_{i,j}^{t}$ signifies the present location of cheetah i $(i = 1, 2, \dots, n)$ in group j $(j = 1, 2, \dots, D)$, n implies the population size, and D denotes the size of the problem. $X_{i,j}^{t+1}i$ signifies the following locations of cheetah i in arrangement j, respectively. t implies the present hunting time, and T is the maximum duration of hunting period. $r_{i,j}^{-1}$ and $\alpha_{i,j}^{t}$ denote the random number and step length for cheetah i in group j.

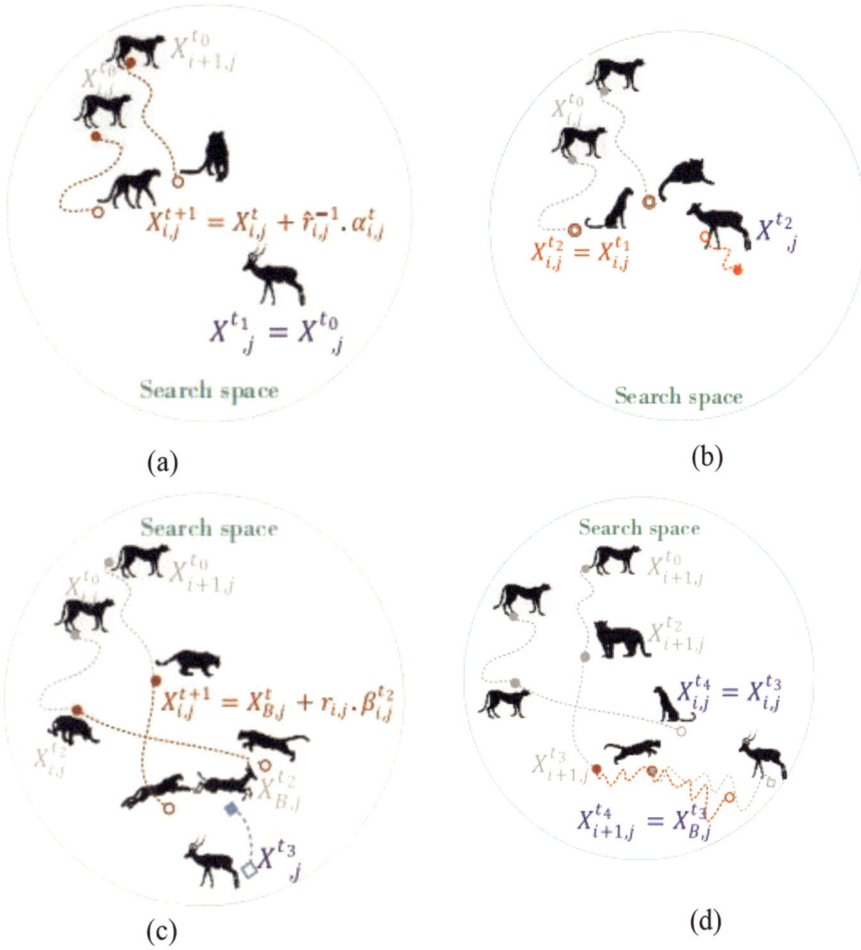

Figure 5.
(a–d) Graphical representation of cheetah's hunting strategies.

3.2.4 Sitting and waiting strategy

The prey might become visible to a cheetah's area of vision, while it is in the searching mode. Each move the cheetah makes in this scenario has the potential to alert the victim to his or her existence and cause the prey to flee. The cheetah may decide to ambush to get sufficiently nearby to the prey in order to allay this worry (by reclining on the ground or lurking amid the bushes) [37]. As a result, the cheetah waits until the prey gets closer while maintaining his or her posture (see **Figure 5b**). The following can be used to simulate this behavior:

$$X_{i,j}^{t+1} = X_{i,j}^{t} \tag{18}$$

where $X_{i,j}^{t+1}$ and $X_{i,j}^{t}$ denote the modified and present locations of cheetah i in arrangement j. In order to boost hunting effectiveness (find a better solution), this method needs the CO approach to refrain from changing all cheetahs concurrently in each group. This can help the algorithm avoid early convergence.

3.2.5 Attacking strategy

A cheetah runs to the prey at maximum speed when it intends to attack. The prey eventually becomes aware of the cheetah's onslaught and starts to flinch. As seen in **Figure 5c**, the cheetah swiftly chases the prey in the line of interception. The cheetah tracks the prey's location and modifies its course so that it blocks the prey's path at a certain point. The prey has to flee and change its location quickly in order stay alive because the cheetah has only traveled a short interval from it at full speed, as illustrated in **Figure 5d**, that is. the cheetah's upcoming location is close to the prey's last position [37]. During this phase, the cheetah captures the prey by moving quickly and maneuvering about. Each cheetah in a group hunt has the ability to change positions depending on the location of the leader or nearby cheetah and the location of the prey. Simply put, all of cheetahs' attacking strategies can be defined numerically as follows:

$$X_{i,j}^{t+1} = X_{B,j}^{t} + r_{i,j} \cdot \beta_{i,j}^{t} \tag{19}$$

where $X_{B,j}^{t}$ denotes the present location of the animal in group j. It is the present optimal location of the population. $r_{i,j}$ and $\beta_{i,j}^{t}$ signify the turning coefficient and interaction coefficient related to the cheetah i in group j.

Based on the hunting behaviors of cheetahs, the proposed CO algorithm incorporates the following assumptions and strategies:

3.2.5.1 Individual representation

In the CO algorithm, each row in the population represents a cheetah in different states. Each column corresponds to a specific arrangement of cheetahs concerning the prey, representing the best solution for each decision variable. Cheetahs emulate the behavior of tracking their prey (the best value for a variable). To identify the optimal solution, cheetahs must successfully capture the prey in each arrangement. A cheetah's performance is assessed through its fitness in all arrangements with higher performance indicating a greater likelihood of successful hunting.

3.2.5.2 Diverse reactions

Just as real cheetahs exhibit different reactions during group hunting, the CO algorithm allows each cheetah to be in various states in each arrangement. Some may be in attack mode while others are in searching, sitting-and-waiting, or attacking modes. Energy levels of cheetahs are considered independent of the prey, and the algorithm introduces random parameters to prevent premature convergence during extensive evolution processes. These random variables act as an energy source for the cheetahs during the hunting process.

3.2.5.3 Random behavior

The behaviors of cheetahs during searching and attacking strategies are assumed to be entirely random, ensuring diversity in the search. In contrast, during the rushing and capturing phases, the prey changes direction abruptly. Randomization parameters model these movements, and varying step lengths and interaction factors with random variables contribute to an effective optimization process.

3.2.5.4 Adaptive strategy

The choice between searching and attacking strategies is random, but searching becomes more likely as a cheetah's energy decreases. Initial steps may be dedicated to searching, while attacking is preferred for larger values of time (t) to achieve better solutions. The selection of strategies is influenced by random factors and energy considerations, much like the cheetah's behavior in the wild.

3.2.5.5 Scanning and sitting-and-waiting

In the CO algorithm, scanning and sitting-and-waiting strategies are considered equivalent, indicating that a cheetah (search agent) remains stationary during the hunting period.

3.2.5.6 Leader adaptation

If the lead cheetah consistently fails in hunting, a randomly selected cheetah's position is changed to the last successful hunting position (i.e., the prey's location). This approach maintains the prey's position among a small population and strengthening the exploration phase.

3.2.5.7 Energy limitations and home range

Each group of cheetahs in the CO algorithm has a time limit for hunting due to energy constraints. If a group fails in a hunting period, they abandon the current prey and return to their home range (initial position). The leader's position is also updated. This strategy helps prevent getting stuck in local optimum solutions.

3.2.5.8 Iterative evolution

In each iteration of the CO algorithm, a subset of the population actively participates in the evolutionary process.

These assumptions and strategies in the CO algorithm draw inspiration from the behavior of cheetahs during hunting, aiming to create an effective optimization technique that mimics their adaptability, randomness, and energy considerations in the quest for optimal solutions.

The fundamental stages of the CO algorithm can be depicted through the pseudo-code outlined in Algorithm 1, drawing inspiration from cheetah hunting tactics and underlying assumptions.

Algorithm 1: The CO methodology

Specify the problem data, dimension (D), and the initial population size (n).
Create the initial population of cheetahs $X_i (i = 1, 2, \ldots, n)$ and assess the fitness of each cheetah.
Set the starting solutions for the population's home, leader, and prey positions.
$t \rightarrow 0$

$it \rightarrow 1$
$MaxIt \rightarrow maximum\ nubmer\ of\ iterations$
$T \rightarrow 60 \times \frac{D}{10}$
while $it \leq MaxIt$
  Randomly select m individuals $(2 \leq m \leq n)$
  for each individual i
  Define its neighbor agent
    for each random arrangement $j \in \{1, 2, \ldots, D\}$
    Calculate $r_{i,j}, \alpha, \beta, H$
    $r_2, r_3 \rightarrow random\ numbers\ between\ [0, 1]$
    $if\ r_2 \leq r_3$
      $r_4 \rightarrow 1 random\ numbers\ between\ [0, 3]$
      $if\ r_4 \leq H$
        Determine the updated position of a member using (19)
      Else
        Determine the updated position of a member using (17)
      End
    Else
        Determine the updated position of a member using (18)
      End
    End
  Revise the solutions of the member and the leader.
End
$t \rightarrow t + 1$
$if\ t > rand \times T$ the leader's position remains unchanged for a certain time, then execute the strategy of abandoning the current prey and returning to the home location, followed by adjusting the leader's position.
End
$it \rightarrow it + 1$
Modify the prey (global best) location
End

4. Results and discussion

The validation process of the introduced WaOA and CO optimization algorithms involved the precise determination of optimal parameters for various PV models. To achieve this, the estimating models, specifically the single-diode model (SDM) and the double-diode model (DDM), were leveraged to compute the PV characteristics, creating power-voltage curves and current-voltage profiles. A comprehensive comparative analysis was undertaken, juxtaposing the estimated performance of each module, both inter-module and against the reference datasheet values of the assessed cells and modules.

The application of the proposed optimization algorithms extended to a range of commercial solar cells and PV modules. This encompassed a standard silicon solar cell from RTC France and the Photo Watt-PWP 201 PV module. Precise measured data from multiple manufacturers' datasheets and references (cited as Refs. [42, 43]) served as the foundational input for these evaluations.

In configuring the optimization process, each of the proposed techniques adhered to a standardized setup. The maximum iteration count was capped at 200 iterations,

	SDM and DDM		PV module	
	Min	**Max**	**Min**	**Max**
I_{ph} (A)	0	1.00	0	2.00
I_{sd} (μA)	0	1.00	0	50.00
R_s (Ω)	0	0.50	0	2.00
R_{sh} (Ω)	0	1000	0	10,000
n_1, n_2, n_3	1.00	2.00	1.00	50.00

Table 1.
Boundary limits for the optimized parameters in solar cell models.

while each population was composed of 20 search agents. The upper and lower limits of the parameters have to be identified are tabulated in **Table 1**. To anchor the entire validation procedure, the MATLAB R2018a platform was utilized, providing a robust and reliable foundation for executing and assessing the optimization algorithms.

In this comprehensive validation procedure, the proposed WaOA and CO optimization algorithms have been rigorously evaluated and benchmarked against established models. The utilization of diverse PV modules, combined with standardized optimization parameters and a trusted computational platform, underscores the meticulous nature of this research effort. The results obtained from this meticulous validation endeavor contribute to the veracity and robustness of the optimization techniques introduced in this study.

4.1 R.T.C. France solar cell

In this study, the researchers employed a novel WaOA and CO optimization technique to identify the parameters of two distinct models of R.T.C. France solar cells. The I–V characteristic curves of the R.T.C. France solar cell were obtained from existing literature sources [41, 42]. To refine the parameters of the SDM (Single-Diode Model) and DDM (Double-Diode Model), the WaOA and CO optimization techniques were employed. The results, including those from the CO-based SDM and WaOA-based SDM models, are tabulated in **Table 2**. This table also encompasses parameter estimates from alternative optimization methods such as An.5-Pt. [44], LW

Technique	I_{ph} (A)	I_{sd} (μA)	R_s (Ω)	R_{sh} (Ω)	n	RMSE
WaOA	0.7607879	0.310682709	0.03654698	52.889880	1.47726717	7.730062E-04
CO	0.760772	0.32384593	0.036366463	53.8160016	1.48143923	7.7912E-04
TGA [16]	0.7606	0.21656	0.0383	50.3996	1.4419	9.90895E-04
COA [15]	0.7607692	0.3083945	0.0365546	52.826661	1.4765477659	7.75467E-04
ABSO [46]	0.76080	0.30623	0.03659	52.2903	1.47583	9.9124E-04
HS [47]	0.7607	0.30495	0.03663	53.5946	1.47538	9.9510E-04
PSO [48]	0.7607	0.400	0.0354	59.012	1.5033	1.3900E-03
GA [49]	0.7619	0.8087	0.0299	42.3729	1.5751	1.8704E-02

Table 2.
Optimized parameters for R.T.C. France solar cell – SDM.

Technique	I_{ph} (A)	I_{sd1} (μA)	I_{sd2} (μA)	R_s (Ω)	R_{sh} (Ω)	n_1	n_2	RMSE
WaOA	0.760814	0.280943	0.25799796	0.03672	53.40962	1.915624	1.4619251	7.631566E-04
CO	0.7607855	0.237803	0.28038851	0.036603	53.75604	1.98260	1.4688138	7.757378E-04
TGA[16]	0.7607	0.16740	0.22083	0.0356	58.2574	1.4999	1.4999	8.48824E-04
COA [15]	0.7607194	0.244676	0.380190	0.03692	53.51296	1.45635	1.989923	7.696186E-04
ABSO[46]	0.76078	0.26713	0.38191	0.03657	54.6219	1.46512	1.98152	9.8344E-04
ABC [53]	0.760813	0.192684	0.999587	0.036861	55.933515	1.438003	1.983721	9.8387E-04
SBMO [54]	0.760786	0.200798	0.74373	0.036917	55.104367	1.441256	1.947888	9.8485E-04
SSO [55]	0.760651	0.287201	0.065979	0.036255	55.853271	1.510345	1.433838	9.9129E-04
MSSO [55]	0.760748	0.234925	0.671593	0.036688	55.714662	1.454255	1.995305	9.8281E-04

Table 3.
Optimized parameters for R.T.C. France solar cell – DDM.

[45], ABSO [46], Newton [50], CM [51], HS [47], PSO [48], GA [49], and PS [52]. Notably, the application of the proposed WaOA algorithm led to the lowest root mean square error (RMSE) value of 7.730062E-04 for the SDM model.

Table 3 further presents outcomes from the utilization of the WaOA and CO techniques to extract parameters from the DDM of the R.T.C. FRANCE solar cell. In order to validate the effectiveness of the employed techniques, this table includes results from other methods such as ABSO [46], HS [47], PSO [48], GA [49], ABC [53], SBMO [54], SSO [55], and MSSO [55]. The results underscore the superior performance of the CO optimization technique, displaying a minimal RMSE objective function value of 7.631566E-04.

Furthermore, the study conducted a comprehensive comparison of the SDM and DDM models utilizing both the WaOA and CO optimization strategies. **Figure 6**

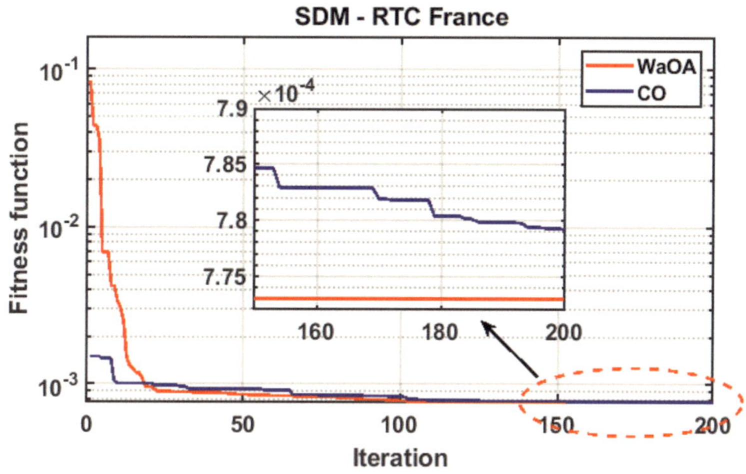

(a)

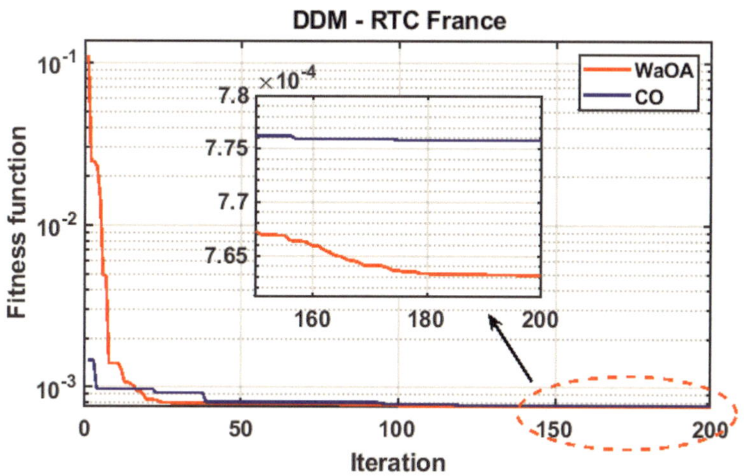

(b)

Figure 6.
Convergence curves for WaOA and CO for RTC France solar cell: (a) SDM and (b) DDM.

graphically illustrates the convergence patterns of the RMSE objective function for both SDM and DDM models of the R.T.C. France solar cell. Notably, the DDM model exhibited faster and more efficient convergence compared to the SDM model. To validate the accuracy of the identified parameters and the efficacy of the proposed optimization methodologies. Moreover, the study depicted the characteristics of the studied solar cell by plotting the estimated parameters against the measured ones for both SDM and DDM models. These plots are presented in **Figures** 7 and **8**, respectively.

To establish the robustness and reliability of the optimization techniques, the researchers executed the optimization algorithms 20 times and recorded the best minimum objective function from each run. Statistical analysis was performed, encompassing metrics such as mean, standard deviation, relative error, as well as best and worst values over the 20 implements. The results of this statistical study are summarized in **Table 4**. Additionally, the final values of the fitness function through the 20 times are graphically presented in **Figure 9**. This analysis convincingly

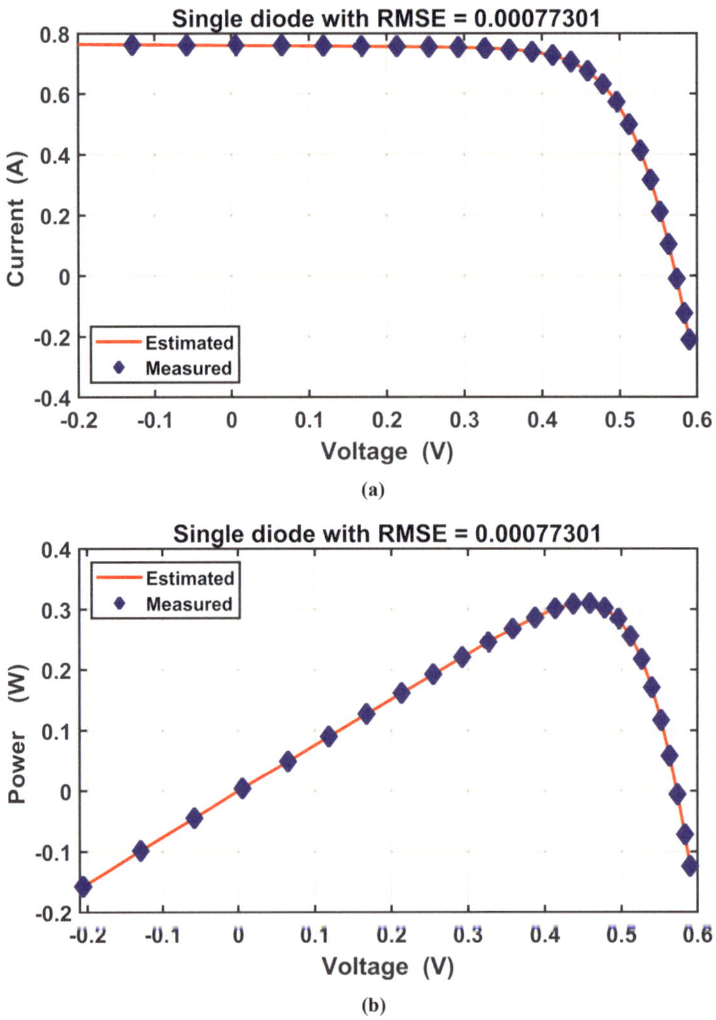

Figure 7.
Characteristics of RTC France solar cell SDM using WaOA: (a) I/V curve and (b) P/V curve.

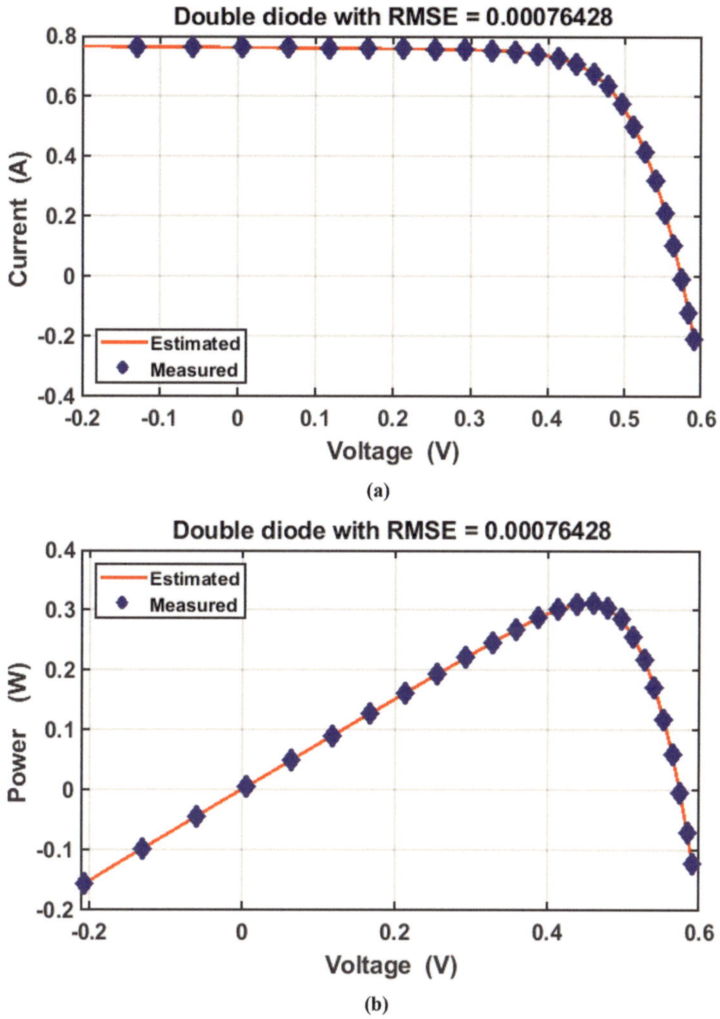

Figure 8.
Characteristics of RTC France solar cell DDM using WaOA: (a) I/V curve and (b) P/V curve.

	SDM		DDM	
Index	**WaOA**	**CO**	**WaOA**	**CO**
Best	0.0007730062	0.0007791220	0.00076315	0.00077573785
Worst	0.000850113	0.0010605183	0.00113209	0.0014436476
Average	0.000785616	0.0008886448	0.00082953	0.0009957970
Median	0.000778313	0.000877633	0.00079522	0.0009532422
STD	0.002131738	0.008997239	0.00928636	0.0200651278
RE	0.326251792	2.81144352	1.73949234	5.67354655
RMSE	2.4304620e-05	0.0001403051	0.000112241	0.000294404

Table 4.
Statistical analysis of R.T.C. France solar cell for both SDM and DDM.

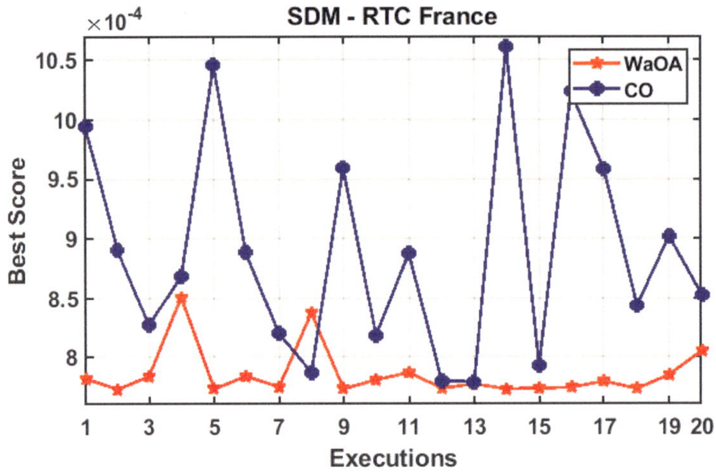

(a)

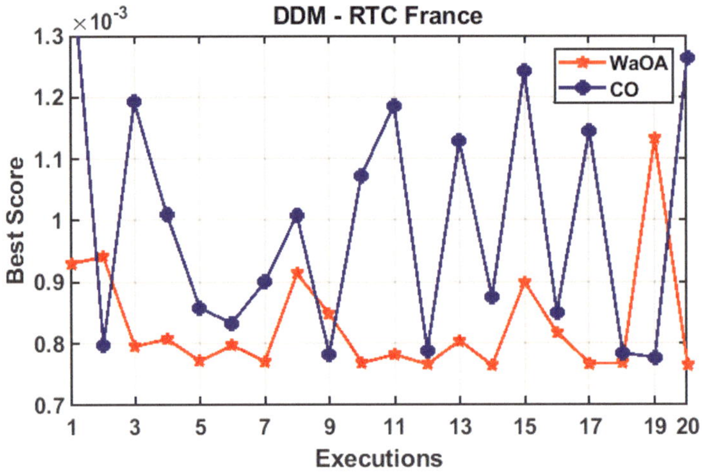

(b)

Figure 9.
Values of RMSE over the 20 runs for WaOA and CO methods for RTC France solar cell: (a) SDM and (b) DDM.

demonstrates that the WaOA algorithm is a highly effective approach for addressing the parameter identification optimization challenges across various mathematical models of the R.T.C. France solar cell.

4.2 Photowatt-PWP201 module

To further validate the effectiveness of the employed WaOA and CO techniques, an assessment was carried out to estimate the parameters for various mathematical models of the Photowatt-PWP201 module. This module comprises 36 silicon in a series combination, operating under conditions of 1000 W/m^2 solar radiation and a cell temperature of 45°C [42, 43]. This chapter aimed to not only assess the proposed methodologies accuracy but also to compare its outcomes with alternative techniques in literature.

The prowess of the proposed algorithms was put to the test in parameter estimation for the SDM concerning the Photowatt-PWP201 module. The findings are detailed in **Table 5**, which additionally featured a comparative analysis against outcomes from other methods, including Newton [56], PS [52], OIS [57], and 1DAB [45]. This comparison effectively demonstrated the COA's superiority in contrast to the other techniques. Notably, the application of WaOA yielded the lowest RMSE value of 0.00212629, indicating its exceptional performance.

To further solidify the validation process, both WaOA and CO algorithms were employed to estimate parameters for the DDM of the Photowatt-PWP201 PV module. The optimized DDM parameters achieved through WaOA and CO are presented in **Table 6**. This table also facilitated a comparison between DDM-based WaOA, DDM-based CO, and other techniques such as WDOWOAPSO [50], GCPSO [58], TVACPSO [59], and ABC-DE [60]. The results highlighted the clear supremacy of the suggested WaOA technique, reflected in the minimized RMSE value of 0.00257992. Importantly, the cumulative results in **Tables 5** and **6** substantiated the WaOA algorithm's efficacy and high precision in parameter extraction for diverse models of the Photowatt-PWP201 module, showcasing a reduction in RMSE value.

Technique	I_{ph} (A)	I_{sd} (μA)	R_s (Ω)	R_{sh} (Ω)	n	RMSE
WaOA	1.0316505	3.287862	1.2076800	872.238281	1.3978671	0.00212629
CO	1.0315416	3.394903	1.20431447	891.945365	1.4013906	0.00225002
TGA[16]	1.0263	9.5710	0.0298	6842.2	1.5255	0.00381949
COA [15]	1.0296281	4.8155440	1.1728971	2000	1.4395994	0.00362202
Newton [56]	1.0318	3.2875	1.2057	555.5556	1.3474016	0.7805
PS [52]	1.0324	3.1859	1.304	843.5233	48.2467	0.0127
OIS [57]	1.03674	3.1946	1.32897	1184.58	49.0435	0.004783
1DAB[45]	1.04276	3.4265	1.73762	948.845	49.2843	0.00536

Table 5.
Optimized parameters for Photowatt-PWP201 module – SDM.

Technique	I_{ph} (A)	I_{sd1} (μA)	I_{sd2} (μA)	R_s (Ω)	R_{sh} (Ω)	n_1	n_2	RMSE
WaOA	1.034168	18.27992	1.13725	1.3265	561.174	3.44090	1.24156	2.5799E-03
CO	1.029743	6.457305	7.33164	1.1241	1587.47	1.42048	47.0336	4.9436E-03
TGA	1.0265	9.2998	2.2586	0.0301	6719.0	1.5225	1.4164	3.7559E-03
COA [15]	1.0265	9.2998	2.2586	1.2163	1019.78	1.34098	50	2.2090E-03
WDOWOAPSO [50]	1.03062	3.171702	5.00	1.2382	744.714	1.31730	1.31730	2.0465E-03
GCPSO [58]	1.032382	2.512916	1.00005	1.2392	744.715	1.31730	1.31693	2.0465E-03
TVACPSO [59]	1.031434	2.638124	1.00	1.2356	821.652	1.32099	2.77777	2.0530E-03
ABC-DE [60]	1.0318	0.32774	2.4305	1.2062	845.249	1.3443	1.3443	2.400E-03

Table 6.
Optimized parameters for Photowatt-PWP201 module – DDM.

The convergence trends of the RMSE based on both WaOA and CO optimization methods across the two proposed models are depicted in **Figure 10**. To visually demonstrate the application of the optimized parameters in the SDM and DDM, estimated I-V (Current-Voltage) and P-V (Power-Voltage) curves of the Photowatt-PWP201 module were generated. These curves, derived from the CO method, were compared with the experimental data, and the results are presented in **Figures 11** and **12** for SDM and DDM, respectively. These figures clearly depicted the alignment between the estimated curves using the CO method and the empirical data.

In order to establish the robustness of the proposed optimization techniques, a statistical analysis was carried out. The final values of the fitness function through the 20 times are graphically presented in **Figure 13**. The outcomes of this analysis were summarized in **Table 7**, demonstrating the CO algorithm's favorable performance

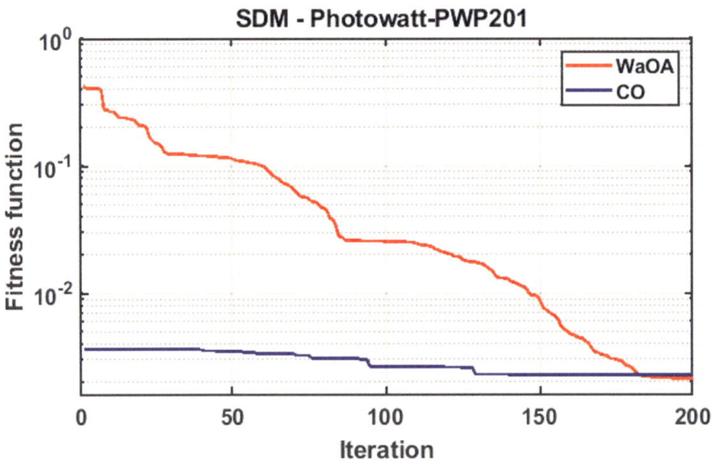

(a)

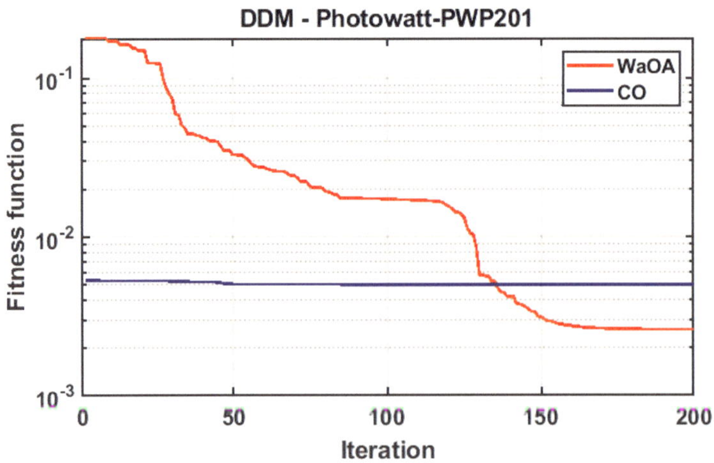

(b)

Figure 10.
Convergence curves for WaOA and CO for Photowatt-PWP201 module: (a) SDM and (b) DDM.

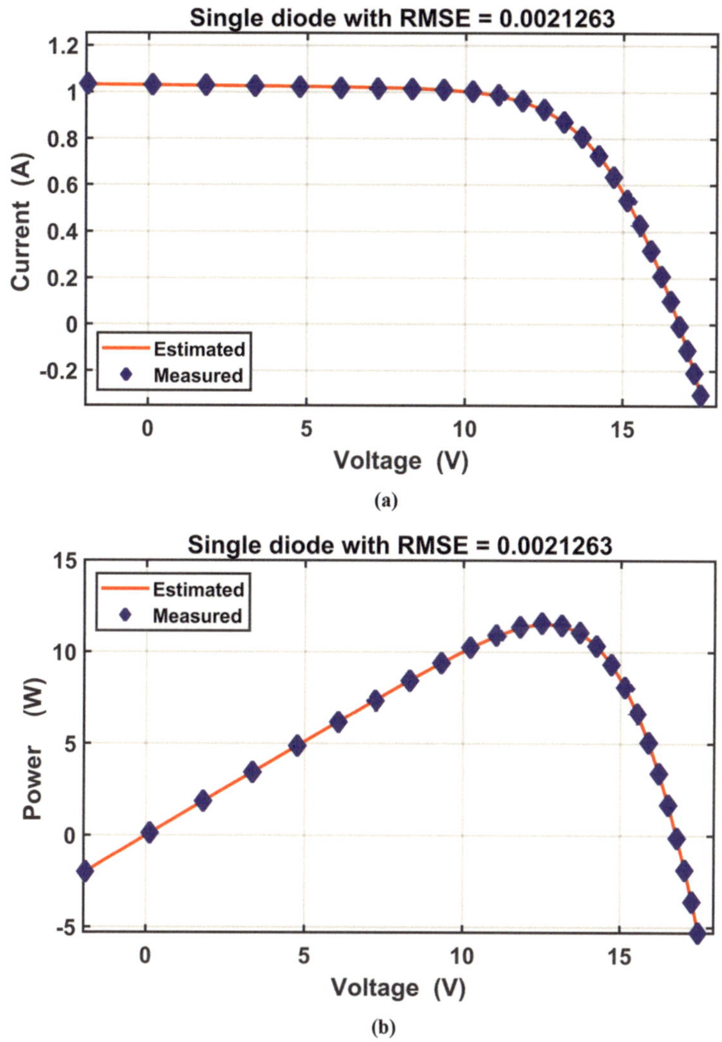

Figure 11.
Characteristics of Photowatt-PWP201 module SDM using WaOA: (a) V/I curve and (b) V/P curve.

with respect to statistical indices such as SD and RE for both estimated models. This comprehensive validation process solidified the CO algorithm's reliability and effectiveness in parameter extraction across different models of the Photowatt-PWP201 PV module.

5. Conclusion

In this chapter, we have harnessed the Walrus Optimization Algorithm (WaOA) and the Cheetah optimizer (CO) to tackle the intricate optimization challenge of determining parameters for solar cells and a variety of PV modules. To thoroughly evaluate the effectiveness of our proposed optimization approach, we have leveraged data from manufacturer datasheets and real-world measurements gathered from

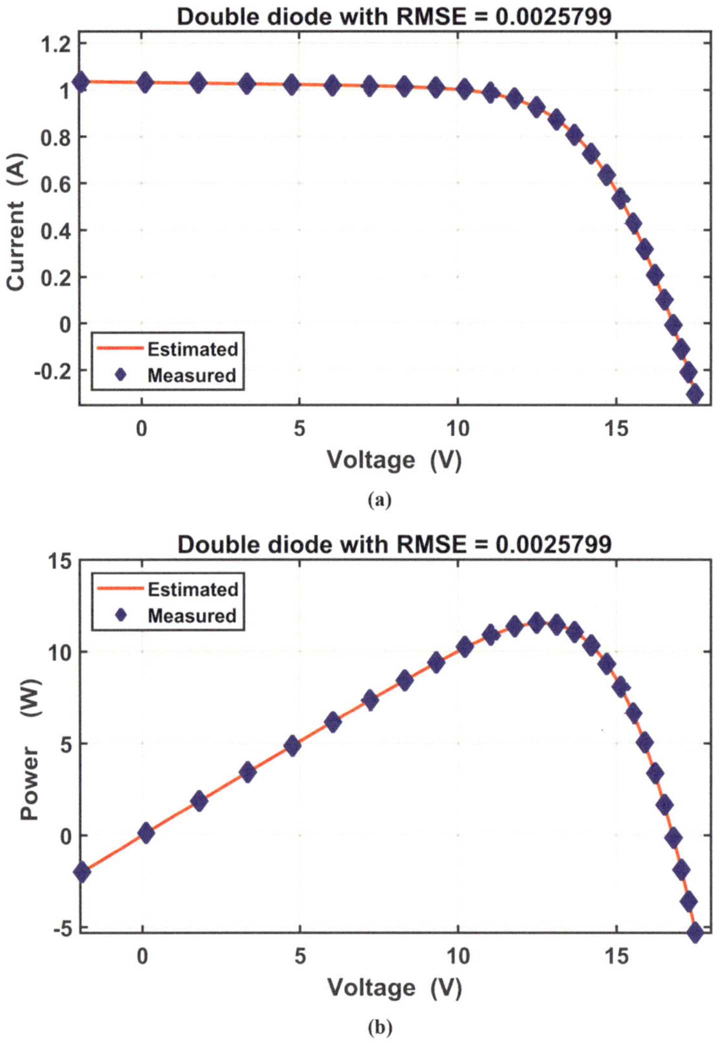

Figure 12.
Characteristics of Photowatt-PWP201 module DDM using WaOA: (a) V/I curve and (b) V/P curve.

literature sources. This comprehensive dataset encompasses diverse solar cells and PV modules, accounting for varying solar radiation intensities and temperatures. Our investigation has encompassed two distinct models—the single-diode model and the double-diode model—for both solar cells and PV modules.

The outcomes resulting from the application of the WaOA and CO have been systematically compared with findings previously documented in the literature pertaining to alternative optimization methodologies. Notably, our proposed algorithms have consistently outperformed others, consistently yielding optimal values for the objective function, often quantified through root mean square error (RMSE). This underscores the robustness and effectiveness of the Walrus Optimization Algorithm and Cheetah optimizer in this context.

Furthermore, we have not limited our assessment to comparative analysis alone; our exploration has delved deeper into the results. By subjecting the outcomes of solar

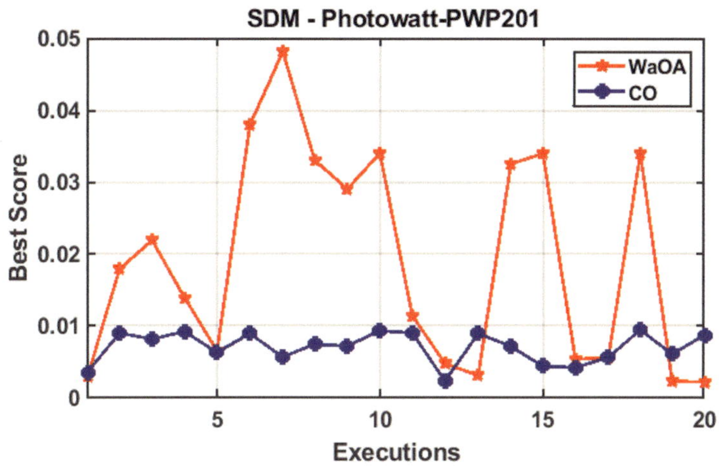

(a)

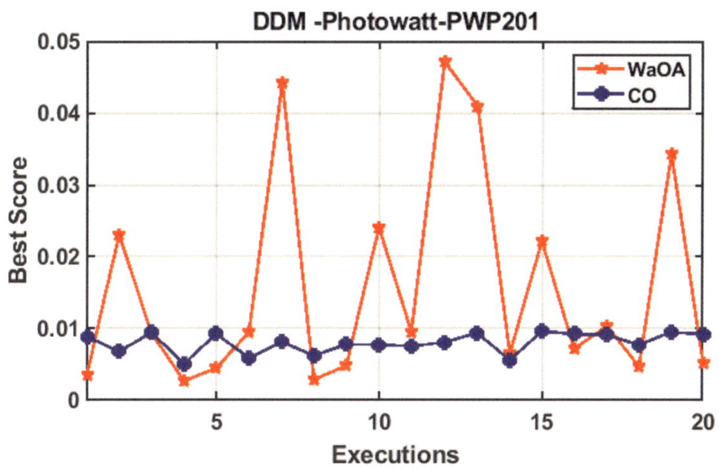

(b)

Figure 13.
Values of RMSE over the 20 runs for WaOA and CO methods for Photowatt-PWP201 module: (a) SDM and (b) DDM.

	SDM		DDM	
Index	**WaOA**	**CO**	**WaOA**	**CO**
Best	0.002126291	0.002250025	0.002579920	0.0049436296
Worst	0.04809182	0.009377722	0.047042367	0.009555128
Average	0.01895375	0.006957418	0.015732435	0.007914192
Median	0.015838395	0.007241249	0.009331317	0.008001565
STD	1.4961108	0.216848518	1.48913560	0.14630791
RE	158.27992	41.84303421	101.9606167	12.01774000
RMSE	0.0222667117	0.005160113	0.019587075	0.003295119

Table 7.
Statistical analysis of Photowatt-PWP201 module for both SDM and DDM.

cell parameter optimization to both parametric and nonparametric statistical scrutiny, we have fortified the confirmation of the efficacy of WaOA and CO in solving this optimization puzzle. A significant achievement has been the remarkable alignment between the V-I curves obtained through the optimized parameters and the corresponding data from manufacturer datasheets. This alignment serves as a clear indication of the validity of our approach.

For instance, when applied to the RTC France solar cell, the WaOA algorithm produced the lowest root mean square error (RMSE) values, measuring at 7.730062E-04 for the SDM model and 7.631566E-04 for the DDM model. Similarly, for the Photowatt-PWP201 module, the WaOA algorithm achieved minimal values for the objective function, measuring at 0.00212629 for the SDM and 0.00257992 for the DDM. These results are a testament to the effectiveness and reliability of our proposed optimization methods.

In summary, the Walrus Optimization Algorithm has proven its mettle by standing alongside established optimization algorithms, firmly establishing itself as a formidable contender for parameter extraction across a diverse range of solar cells and PV modules. Through rigorous testing and analysis, we have substantiated the potential of our methodology to make a significant contribution to the ongoing advancement of solar technology optimization.

For future directions, it is worth considering the application of these optimization algorithms to more complex solar systems, exploring potential enhancements in convergence speed, and adapting the approach to real time, dynamic operational scenarios for solar cells and PV modules. Additionally, the incorporation of machine learning and artificial intelligence techniques for predictive modeling and optimization could further advance the state of the art in solar technology.

Author details

Hamdy M. Sultan[1], Mahmoud A. Mossa[1] and Almoataz Y. Abdelaziz[2]*

1 Faculty of Engineering, Electrical Engineering Department, Minia University, Minia, Egypt

2 Faculty of Engineering, Ain Shams University, Cairo, Egypt

*Address all correspondence to: almoataz_abdelaziz@eng.asu.edu.eg

References

[1] Ridha HM, Heidari AA, Wang M, Chen H. Boosted mutation-based Harris hawks optimizer for parameters identification of single-diode solar cell models. Energy Conversion and Management. 2020;**209**:112660

[2] Li Y, Chiu Y, Lin T-Y. Research on new and traditional energy sources in OECD countries. International Journal of Environmental Research and Public Health. 2019;**16**(7):1122

[3] Xiong G, Li L, Mohamed AW, Yuan X, Zhang J. A new method for parameter extraction of solar photovoltaic models using gaining–sharing knowledge based algorithm. Energy Reports. 2021;7:3286-3301

[4] Diab AAZ, Sultan HM, Kuznetsov ON. Optimal sizing of hybrid solar/wind/hydroelectric pumped storage energy system in Egypt based on different meta-heuristic techniques. Environmental Science and Pollution Research. 2020;**27**: 32318-32340

[5] Saad M, Sultan HM, Abdeltwab A, Diab AAZ. Feasibility study of a small-scale grid-connected PV power plants in Egypt; case study: New Valley governorate. International Journal of Advanced Computer Science and Applications. 2021;**12**(6)

[6] Sultan HM, Menesy AS, Kamel S, Alghamdi AS, Zohdy M. Optimal sizing of isolated hybrid PV/WT/FC system using Manta ray foraging optimization algorithm. International Transactions on Journal of Engineering and Management Applied Science and Technology. 2020; **11**:1-12

[7] Reinders A, Verlinden P, Van Sark W, Freundlich A. Photovoltaic Solar Energy: From Fundamentals to Applications. New Jersey, USA: John Wiley & Sons; 2017

[8] Rojas D, Rivera M, Wheeler P. Basic principles of solar energy. In: 2021 IEEE CHILEAN Conference on Electrical, Electronics Engineering, Information and Communication Technologies (CHILECON). Valparaíso, Chile: IEEE; 2021. pp. 1-6

[9] Chee AK. The mechanistic determination of doping contrast from Fermi level pinned surfaces in the scanning electron microscope using energy-filtered imaging and calculated potential distributions. Microscopy and Microanalysis. 2022; **28**(5):1538-1549

[10] Chee AK, Broom RF, Humphreys CJ, Bosch EG. A quantitative model for doping contrast in the scanning electron microscope using calculated potential distributions and Monte Carlo simulations. Journal of Applied Physics. 2011;**109**(1):1-9

[11] Chee AK. Unravelling new principles of site-selective doping contrast in the dual-beam focused ion beam/scanning electron microscope. Ultramicroscopy. 2020;**213**:112947

[12] Kamel OM, Diab AAZ, Mahmoud MM, Al-Sumaiti AS, Sultan HM. Performance enhancement of an islanded microgrid with the support of electrical vehicle and STATCOM systems. Energies. 2023; **16**(4):1577

[13] Diab AAZ, Mohamed MA, Al-Sumaiti A, Sultan H, Mossa M. A novel hybrid optimization algorithm for maximum power point tracking of partially shaded photovoltaic systems.

In: Advanced Technologies for Solar Photovoltaics Energy Systems. Berlin, Germany: Springer; 2021. pp. 201-230

[14] Abdelghany RY, Kamel S, Sultan HM, Khorasy A, Elsayed SK, Ahmed M. Development of an improved bonobo optimizer and its application for solar cell parameter estimation. Sustainability. 2021;**13**(7):3863

[15] Abdelghany RY, Kamel S, Ramadan A, Sultan HM, Rahmann C. Solar cell parameter estimation using school-based optimization algorithm. In: 2021 IEEE International Conference on Automation/XXIV Congress of the Chilean Association of Automatic Control (ICA-ACCA). Valparaíso, Chile: IEEE; 2021. pp. 1-6

[16] Diab AAZ, Sultan HM, Do TD, Kamel OM, Mossa MA. Coyote optimization algorithm for parameters estimation of various models of solar cells and PV modules. IEEE Access. 2020;**8**:111102-111140

[17] Diab AAZ, Sultan HM, Aljendy R, Al-Sumaiti AS, Shoyama M, Ali ZM. Tree growth based optimization algorithm for parameter extraction of different models of photovoltaic cells and modules. IEEE Access. 2020;**8**: 119668-119687

[18] Humada AM et al. Modeling of PV system and parameter extraction based on experimental data: Review and investigation. Solar Energy. 2020;**199**: 742-760

[19] Pillai DS, Rajasekar N. Metaheuristic algorithms for PV parameter identification: A comprehensive review with an application to threshold setting for fault detection in PV systems. Renewable and Sustainable Energy Reviews. 2018;**82**:3503-3525

[20] Xiong G, Zhang J, Shi D, Zhu L, Yuan X, Yao G. Modified search strategies assisted crossover whale optimization algorithm with selection operator for parameter extraction of solar photovoltaic models. Remote Sensing. 2019;**11**(23):2795

[21] Abbassi R, Abbassi A, Jemli M, Chebbi S. Identification of unknown parameters of solar cell models: A comprehensive overview of available approaches. Renewable and Sustainable Energy Reviews. 2018;**90**:453-474

[22] Senturk A, Eke R. A new method to simulate photovoltaic performance of crystalline silicon photovoltaic modules based on datasheet values. Renewable Energy. 2017;**103**:58-69

[23] Gnetchejo PJ, Essiane SN, Ele P, Wamkeue R, Wapet DM, Ngoffe SP. Enhanced vibrating particles system algorithm for parameters estimation of photovoltaic system. Journal of Power and Energy Engineering. 2019;**7**(08):1

[24] Tossa AK, Soro Y, Azoumah Y, Yamegueu D. A new approach to estimate the performance and energy productivity of photovoltaic modules in real operating conditions. Solar Energy. 2014;**110**:543-560

[25] Ghani F, Fernandez E, Almonacid F, O'Donovan T. The numerical computation of lumped parameter values using the multi-dimensional Newton-Raphson method for the characterisation of a multi-junction CPV module using the five-parameter approach. Solar Energy. 2017;**149**: 302-313

[26] Yang B et al. Comprehensive overview of meta-heuristic algorithm applications on PV cell parameter identification. Energy Conversion and Management. 2020;**208**:112595

[27] Oliva D, Abd El Aziz M, Hassanien AE. Parameter estimation of photovoltaic cells using an improved chaotic whale optimization algorithm. Applied Energy. 2017;**200**:141-154

[28] Mughal MA, Ma Q, Xiao C. Photovoltaic cell parameter estimation using hybrid particle swarm optimization and simulated annealing. Energies. 2017;**10**(8):1213

[29] Gong L, Cao W, Zhao J. An improved PSO algorithm for high accurate parameter identification of PV model. In: 2017 IEEE International Conference on Environment and Electrical Engineering and 2017 IEEE Industrial and Commercial Power Systems Europe (EEEIC/I&CPS Europe). Milan, Italy: IEEE; 2017. pp. 1-5

[30] Fathy A, Rezk H. Parameter estimation of photovoltaic system using imperialist competitive algorithm. Renewable Energy. 2017;**111**:307-320

[31] Ram JP, Babu TS, Dragicevic T, Rajasekar N. A new hybrid bee pollinator flower pollination algorithm for solar PV parameter estimation. Energy Conversion and Management. 2017;**135**: 463-476

[32] Babu TS, Ram JP, Sangeetha K, Laudani A, Rajasekar N. Parameter extraction of two diode solar PV model using fireworks algorithm. Solar Energy. 2016;**140**:265-276

[33] Rezk H, Fathy A. A novel optimal parameters identification of triple-junction solar cell based on a recently meta-heuristic water cycle algorithm. Solar Energy. 2017;**157**:778-791

[34] Abdelghany RY, Kamel S, Sultan HM, Hassan MH, Nasrat L. Optimal solar cell parameter estimation based on sooty tern optimization algorithm. In: 2022 23rd International Middle East Power Systems Conference (MEPCON). Cairo, Egypt: IEEE; 2022. pp. 1-6

[35] Trojovský P, Dehghani M. A new bio-inspired metaheuristic algorithm for solving optimization problems based on walruses behavior. Scientific Reports. 2023;**13**(1):8775

[36] Trojovský P, Dehghani M. Walrus Optimization Algorithm: A New Bio-Inspired Metaheuristic Algorithm. Florida, USA; 2022

[37] Akbari MA, Zare M, Azizipanah-Abarghooee R, Mirjalili S, Deriche M. The cheetah optimizer: A nature-inspired metaheuristic algorithm for large-scale optimization problems. Scientific Reports. 2022;**12**(1):10953

[38] Gupta J, Nijhawan P, Ganguli S. Parameter estimation of different solar cells using a novel swarm intelligence technique. Soft Computing. 2022;**26**(12): 5833-5863

[39] Ayyarao TS, Kumar PP. Parameter estimation of solar PV models with a new proposed war strategy optimization algorithm. International Journal of Energy Research. 2022;**46**(6): 7215-7238

[40] Rawa M et al. Estimation of parameters of different equivalent circuit models of solar cells and various photovoltaic modules using hybrid variants of honey badger algorithm and artificial gorilla troops optimizer. Mathematics. 2022;**10**(7):1057

[41] Chee AK. Quantitative dopant profiling by energy filtering in the scanning electron microscope. IEEE Transactions on Device and Materials Reliability. 2016;**16**(2):138-148

[42] Brinciotti E et al. Frequency analysis of dopant profiling and capacitance spectroscopy using scanning microwave microscopy. In: IEEE Transactions on Nanotechnology. Vol. 16, No. 1. Jan 2017. pp. 75-82

[43] Chee AK. Enhancing doping contrast and optimising quantification in the scanning electron microscope by surface treatment and Fermi level pinning. Scientific Reports. 2018;**8**:5247

[44] Chee AK. Fermi level pinning characterisation on ammonium fluoride-treated surfaces of silicon by energy-filtered doping contrast in the scanning electron microscope. Scientific Reports. 2016;**6**(1):32003

[45] Chee AK. Principles of high-resolution dopant profiling in the scanning helium ion microscope, image widths, and surface band bending. IEEE Transactions on Electron Devices. 2019; **66**(11):4883-4887

[46] Yu S, Heidari AA, Liang G, Chen C, Chen H, Shao Q. Solar photovoltaic model parameter estimation based on orthogonally-adapted gradient-based optimization. Optik. 2022;**252**:168513

[47] Niu Q, Zhang H, Li K. An improved TLBO with elite strategy for parameters identification of PEM fuel cell and solar cell models. International Journal of Hydrogen Energy. 2014;**39**(8):3837-3854

[48] Bouzidi K, Chegaar M, Aillerie M. Solar cells parameters evaluation from dark IV characteristics. Energy Procedia (UK). 2012;**18**:1601-1610

[49] Askarzadeh A, Rezazadeh A. Artificial bee swarm optimization algorithm for parameters identification of solar cell models. Applied Energy. 2013;**102**:943-949

[50] Zhou J, Zhang Y, Zhang Y, Shang W-L, Yang Z, Feng W. Parameters identification of photovoltaic models using a differential evolution algorithm based on elite and obsolete dynamic learning. Applied Energy. 2022;**314**: 118877

[51] Jiang J, Xu L, Song C, Bao H. An improved ABC algorithm and its application on parameter identification of solar cell model. In: 15th International Symposium on Computational Intelligence and Design (ISCID). IEEE; 2022. pp. 236-240

[52] Nunes H, Pombo J, Bento P, Mariano S, Calado M. Collaborative swarm intelligence to estimate PV parameters. Energy Conversion and Management. 2019;**185**:866-890

[53] Mares O, Paulescu M, Badescu V. A simple but accurate procedure for solving the five-parameter model. Energy Conversion and Management. 2015;**105**:139-148

[54] Askarzadeh A, Rezazadeh A. Parameter identification for solar cell models using harmony search-based algorithms. Solar Energy. 2012;**86**(11): 3241-3249

[55] Ye M, Wang X, Xu Y. Parameter extraction of solar cells using particle swarm optimization. Journal of Applied Physics. 2009;**105**(9)

[56] AlRashidi M, AlHajri M, El-Naggar K, Al-Othman A. A new estimation approach for determining the I–V characteristics of solar cells. Solar Energy. 2011;**85**(7):1543-1550

[57] AlHajri M, El-Naggar K, AlRashidi M, Al-Othman A. Optimal extraction of solar cell parameters using pattern search. Renewable Energy. 2012; **44**:238-245

[58] Oliva D, Cuevas E, Pajares G. Parameter identification of solar cells using artificial bee colony optimization. Energy. 2014;**72**:93-102

[59] Yu K, Liang J, Qu B, Chen X, Wang H. Parameters identification of photovoltaic models using an improved JAYA optimization algorithm. Energy Conversion and Management. 2017;**150**: 742-753

[60] Lin P, Cheng S, Yeh W, Chen Z, Wu L. Parameters extraction of solar cell models using a modified simplified swarm optimization algorithm. Solar Energy. 2017;**144**:594-603

Chapter 4

Use of Concentrated Solar Power Technology for a High Temperature Processes: Case Study of Uzbekistan

Mukhammad-Sultan Payzullakhanov, Rasul Akbarov and Lola Suvonova

Abstract

The paper examines the state and prospects for the development of renewable energy use in Uzbekistan, presents the specific features and conditions of concentrated solar power (CSP) technology, analyzes the technological capabilities of high-temperature solar furnaces as one of the promising areas of CSP technology, and notes specific scientific directions in this area. A comparative analysis of the technical and optical-energy characteristics of high-temperature solar furnaces of well-known research centers and their application in solving of actual scientific and technical problems is carried out. The main parameters and energy characteristics of the Uzbek 1000 kWt solar furnace in Parkent (BSF) in comparison with the similar French (Odeillo) furnace are given. Special equipment and stands installed in the focal zone of the furnace for melting, synthesis, heat treatment and complex testing of various technical products in high-temperature conditions are presented. By the example of the analysis of the process of quenching high-temperature materials, some features of high-temperature processes are shown.

Keywords: Solar furnace, concentrator, heliostat, mirrors, focal zone, focal length, energy distribution

1. Introduction

1.1 Renewable energy sources in Uzbekistan

Currently, the use of renewable energy sources (RES) is a priority in most countries of the world, including Uzbekistan. Such a global situation is caused by many well-known objective factors, the main of which are: depletion of non-renewable energy resources of the Earth, colossal emission of carbon dioxide into the atmosphere, uneven distribution of fossil energy resources and different natural, geographical and climatic conditions of the countries of the world (the number of sunny days, the presence of rivers, etc.), political and economic conjuncture and others.

Therefore, the intensive transition to the use of renewable energy sources is an important stage in the development of global energy [1–5].

In Uzbekistan, despite the high potential of, for example, solar energy, major projects on the use of alternative energy sources have not been implemented until recently. In 2020, the total installed capacity of renewable energy plants was 2 gigawatts and almost 100% of this capacity was accounted for by hydroelectric power plants [6]. In recent years, the situation has changed and now Uzbekistan pays special attention to the use of renewable energy [4, 7, 8]. A number of important laws on renewable energy have been adopted, an attractive investment climate has been formed, etc. In 2021, a photovoltaic power plant with a capacity of 100 MW was commissioned in Navoi region and in 2022 a similar station in the Samarkand region. This year, investment agreements were signed between the Ministry of Investments of Uzbekistan and ACWA Power, which provides for the commissioning of two solar photovoltaic plants with a total capacity of 1400 MW in the Tashkent and Samarkand regions and three energy storage systems with a total capacity of 1200 MW. It should be noted that by 2030 it is planned to produce 35% of electricity from renewable energy sources. To do this, 15,000 MW of capacity will be put into operation – 10,000 MW of solar power plants and 5000 MW of wind power plants [9].

At the same time, it should be noted that the introduction of small-scale developments (in power from several hundred watts to several hundred kilowatts) for renewable energy, mainly for photovoltaic installations and thermal collectors, in various organizations, in the agro-industrial sector, in social facilities, in individual and rural farms and etc. is developing at an accelerated step [10].

1.2 Concentrated solar power technology

One of the promising areas of renewable energy use is the technology of concentrated solar power (CSP). This category mainly includes solar point-focus concentrators (Dish), solar linear-focus concentrators (PTC, Fresnel) and tower-type solar stations. It should be noted that these installations, in addition to wide practical applications, in most cases are also unique scientific tools for conducting research in the field of high-temperature processes. The disadvantages of such systems include the high complexity of their installation and the high cost compared to photovoltaic systems. In addition, concentrating installations operate from direct solar radiation and therefore it is almost mandatory to have a tracking system for the Sun and therefore regions with a predominance of direct solar radiation are preferable for such tasks.

Despite these disadvantages of CSP technologies, due to the possibility of this technology to achieve high temperatures, up to 3000 degrees, CSP technologies have found wide applications in the world. For example, as noted in [11], CSP is a highly competitive technology for generating electricity, research in the field of solar materials science is known and relevant, in particular the successes of scientists from France and Uzbekistan [12–20]. The paper [20] provides a brief overview of the dynamics of the development of solar energy concentration technologies in the world in the period from 2010 to 2021 and assesses the possibility of using these technologies in Uzbekistan. It should be noted that the climatic conditions of Uzbekistan correspond to such and other necessary criteria [21, 22].

It should be noted that the global generating potential of CSP is estimated at about a billion terawatt-hours (TWh) per year, which is tens of thousands of times higher than global electricity consumption [23]. It is predicted that by 2050, 12 and 25% of the total and electric energy consumed by mankind, respectively, will be generated using CSE technology [24]. Currently, linear-focus concentrators, which generate up to 80% of the energy generated using CSE technology, have found the greatest use among CSP-based installations [25].

In recent years, fundamental and scientific and technical research, as well as applied developments using CSP technology, have been actively carried out. Currently, there are several tens of thousands of solar installations for various purposes and capacities based on CSP technology in the world. There are about 100 large stations based on CSP technology in the world, and about the same number of such stations are planned to be built. The undisputed leaders in the use of CSP, especially for large projects, are Spain and the USA. In Australia, in the countries of North Africa, in Chile, Algeria, Morocco, Egypt, CSP technologies are also widely used. Recently, China, France, Germany, India, Israel, Italy and South Korea have been increasingly working in this direction. By the end of 2019, the shares of installed capacities of stations based on CSE by countries of the world were: Spain (37%), USA (28%), China (10), Morocco (8), South Africa (8), Israel (6) and the rest of the countries 3% [26].

1.3 Reasoning of the research directions

An analysis of the trends in the development of alternative energy sources shows that many countries are already paying special attention to these energy sources. The reasons for this, as already noted, are obvious - the lack of energy due to the growth of its consumption, the depletion of fossil energy sources, and serious global environmental problems. The way out of this situation is the uses of energy-saving and environmentally friendly energy production technologies, that is, Renewable Energy Sources (RES).

Here the most important thing is a reasonable understanding of this energy transition in each country and the importance of this step, the country's contribution to solving global energy problems in preventing global climate change on our planet.

It should be noted that alternative energy sources are widely classified according to the types of their origin - solar energy, wind energy, geothermal energy, hydropower, ocean energy, bioenergy and others. At the same time, if we consider, for example, the total generation of electricity in the world, the share of RES is 27.3% (end of 2019, [4]). And here, 15.9% hydro, 5.9% wind, 2.8% solar panels, 2.2 biofuels and the rest 0.4% are geothermal, CSP and ocean energy.

This paper discusses the use of concentrated solar energy in the study of some high-temperature processes in the context of the use and presentation of solar technologies.

The main purpose of this work is to present the technological capabilities of the BSF in Parkent in the study of the above problems, as well as a description of the accompanying series of technological equipment. Before directly presenting the main material, we briefly presented the state and prospects of the uses of renewable energy sources in Uzbekistan and a brief overview of the technology of concentrated solar power (CSP). It would seem that the questions are far from the main topic.

However, if you look closely, these areas are interconnected, these materials complement each other, allow a more comprehensive understanding of the specific task under consideration and see its perspective. It should be noted that during the construction of the BSF (80s), the state and prospects of CSP technologies were analyzed and also, of course, local solar energy resources were studied, etc.

The reason for the analysis is the efficiency of the BSF, that is the results of high-temperature studies, along with other achievements, ultimately depend on the level of direct solar radiation, its daily and annual duration, and so on. For this reason, a modern meteorological station was built at the BSF location.

Thus, the studies presented in this paper make it possible to get acquainted with the unique technological capabilities of the BSF, the materials provide information on some well-known high-temperature solar furnaces, promote the development of research in the field of high-temperature processes and the development of cooperation with Uzbek scientists.

2. High temperature solar furnaces

As you know, to get 1 MWh of energy, you need to burn 250 kilograms of coal at a consumption of 0.67 tons of oxygen, with the formation of 0.9 tons of carbon dioxide, polluting the environment. The constant increase in energy consumption, depletion of fossil fuel reserves and environmental problems arising from its combustion pose the task of the energy economy of the Republic of Uzbekistan to search and develop new non-traditional energy sources, for example, renewable and environmentally friendly solar energy. There are 260 sunny days a year in Uzbekistan and the pace of development of solar technologies can become a determining factor in the development of savings.

High temperatures can be obtained, in particular, in solar high-temperature installations - solar concentrators. At the same time, the most suitable solar installations are point-focus concentrators.

The geometric shape of the reflecting surface of such concentrators is usually formed by the rotation of conic sections - spheres, parabolas and hyperbolas. Examples of such concentrators are: spherical concentrator, paraboloid concentrator (Dish system), parabolic trough concentrator (PTC), hyperboloid concentrator (Cassegrain optical system) and others. At the same time, when designing such installations, in order to facilitate the technological processes of forming reflective surfaces of installations, various simplified technical solutions are used. An example of such approaches is the concentrators with facets, the use of Fresnel reflectors, the use of approximately close to ideal geometric shapes, and others. As shown in [27], paraboloid concentrators have the highest concentrating capacity.

In the aspect of the use of solar radiation, the most technologically effective is its concentration on the surface by means of mirror concentrating systems (MCS). To date, there are many MCS of various sizes and the level of concentration of solar radiation. Among high-temperature solar furnaces, first of all, it is necessary to note the unique large poly-heliostat solar furnaces with a capacity of 1000 kW (**Figure 1**) in Parkent (Uzbekistan, 1987) and in Odeillo (France, 1971) [28, 29]. It has become generally accepted that the abbreviation BSF - Big Solar Furnace is used to identify these furnaces. The diameters of the concentrators of these furnaces are 54 m with a focal length of 18 m. The density distribution of concentrated solar energy in the focal zone of these furnaces has the form of a Gaussian distribution with a diameter of almost 1 m and a concentration of about 10,000 times. In these furnaces, in the best condition of their optical elements, a temperature of about 3000 degrees is reached. The maximum energy density reaches up to 750 Wt/cm^2, the spot diameter is about 1 m. Such densities of the concentrated solar radiation flux make it possible to implement high-temperature physico-chemical processes leading to the synthesis of new materials with a special microstructure, the production of hydrogen by thermochemical method, the generation of electrical and thermal energy, laser radiation and etc. In

Figure 1.
General view of BSF in Parkent and Odeillo.

addition, it should be noted that the high accuracy of the optical elements of the BSF allows it to be used as a multifunctional ground-based Cherenkov telescope for recording Cherenkov flashes from broad atmospheric showers of cosmic rays with an energy of $E_0 = 10^{13} - 10^{15}$ eV.

Thus, it can be stated that BSF, in addition to being an environmentally friendly melting furnace, is also a unique research tool for conducting high-temperature studies and has a number of important significant advantages compared to other high-temperature solar furnaces, and they are as follows:

1. High furnace capacity, ≈ up to 1 MW.

2. High levels of the coefficient of concentration of solar radiation, 4500÷10,000.

3. Large size of the focal spot of the furnace, ≈ 80÷100 cm.

4. The presence of 62 heliostats, which allows manipulating the focal distribution of energy.

5. The presence of automatic control system of heliostats (ACS), which allows you to flexibly control the movements of heliostats.

6. Simulation of 62 narrow-aperture concentrators.

7. The ability to control the concentrated solar energy flux in technological processes.

8. The ability to measure and control the optical-energy characteristics of the furnace, etc.

The Institute of Materials Science has developed and created various small solar furnaces (SSF) with an average capacity of 1500 Wt and a couple of them have been exported to Egypt (Tabbin Institute of Metallurgy, TIMS, Cairo) and India (International Powder Metallurgy Center, ARCI, Hyderabad). The exported furnaces are identical to each other and have a thermal power of 1500 watts. The installation consists of a single flat heliostat with size of 2.8 × 2.8 m, a single paraboloid concentrator with a diameter of 2 m, a solar sensor and auxiliary measuring instruments and the necessary equipment for its operation. The heliostat of the installation has an automated system for tracking of the Sun. These furnaces are shown in the **Figure 2**.

Currently, fundamental and applied research in the field of solar materials science using high-temperature solar furnaces is carried out in many leading scientific centers of the world. They are, for example, NREL (USA), Sandia National Laboratory, USA, PROMES (France), Plataforma Solar de Almernia (PSA, Spain) ETH Zurich (Switzerland), Paul Scherrer Institute (Switzerland), DLR (Germany), The Weizmann Institute of Science (Israel), HRFSF (Mexico), Masdar Institute Solar Platform (Abu Dhabi, UAE), Institute of Materials Science (Uzbekistan) and many others. It should be noted that [30] presents a fairly detailed analysis of the characteristics of various solar furnaces used for the production of "green" hydrogen, in particular, the possibilities of the Parkent solar furnace are also noted there. In the well-known work [31], a number of important studies in the field of high-temperature processes performed using solar furnaces are presented in sufficient detail - solar surface hardening of steel, refinement of nanomaterials, solar synthesis of fullerenes and carbon nanotubes.

3. Measurement of highly concentrated solar flux

It should be noted that any complex technological installation without a system for monitoring and measuring the relevant parameters and its performance characteristics cannot be operated efficiently and reliably. In this regard, a Big Solar Furnace is equipped with modern devices for measuring the density of concentrated solar

Figure 2.
Installed solar furnaces in Egypt and the Republic of India.

energy, the temperature of the materials under study and devices for monitoring the flow of high-temperature processes. These include a high-temperature pyrometer for remote measurement of the temperature of materials (IMPAC IGA 12, Tmax = 3500° C), a thermal imager (FLIR A655), various radiometers, photometers, digital thermometers, a vision system (developed by the Institute of Materials Science), etc.

Thus, one of the important problems in the operation of high-temperature solar installations is the measurement of the flux density in the focal zone of the concentrator. For this purpose, calorimetric, radiometric and photometric methods are widely used, as well as television measuring systems.

The most common method of measuring the energy flux density in the focal plane of the Mirror Concentrating System (MCS) is the calorimetric method, based on measuring the amount of heat transferred by the surface of the calorimeter to the coolant. The calorimetric method consists in measuring the amount of energy transferred from the MCS into the beam-receiving cavity at different diameters of the inlet holes (**Figure 3**).

Special diaphragms are installed tightly to the inlet of the radiating cavity so that the center of the orifice of the diaphragms coincides with the center of the orifice of the radiating cavity. The main advantage of the calorimetric method is to obtain an absolute value of the energy distribution, as well as the ability to measure radiant fluxes in a wide range of energy densities up to 1000 Wt/cm^2, with a relative error of 10–15%.

The radiometric measurement method, like the calorimetric one, is based on the thermal effect of radiant energy, however, the flow density in this case is judged directly by measuring the thermo-EMF or thermal resistance of the receiving surface. Modern designs of radiometers with cooling of the measuring part make it possible to measure consistently high levels of flux density up to $1.5*10^7$ Wt/m^2. However, radiometers measure only relative values of the flux density.

One of the modern and informative methods of measuring the values of the flux density is the so-called system of technical view-STV. In this method, the irradiance is determined based on the image of the focal spot. The obtained relative values are then converted to absolute values using calibration coefficients. STV allows you to instantly get all the information about the focal spot. The obtained data is easily processed using graphic editors on personal computers.

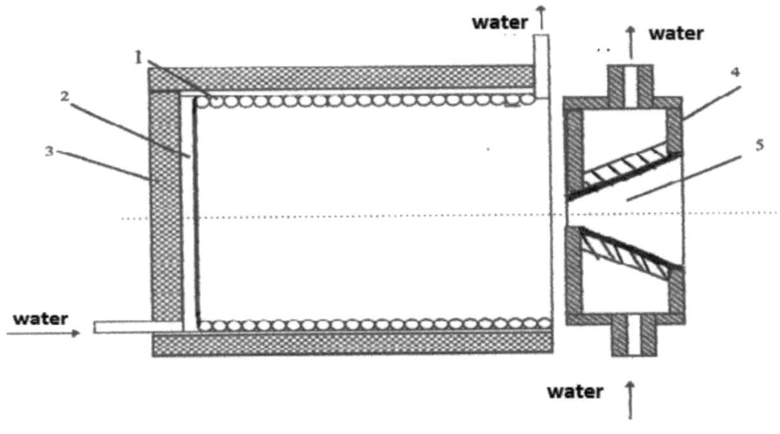

Figure 3.
Design of the flow calorimeter. 1-thin-walled copper tubes; 2-sheet copper; 3-thermal insulation; 4-diaphragm; 5-inlet.

Pyrometers operating on the basis of the laws of thermal radiation of heated bodies are widely used to measure high temperatures. It is known that the flow of radiant energy (Q) falling on the surface of the body is partially reflected (R), partially passes (D), and the rest is absorbed (A), and Q = R + D + A. The absorbed energy is converted into the energy of the thermal motion of the molecules. The material absorbing the rays heats up and emits radiant energy in the form of electromagnetic waves of various lengths, the intensity of which increases with increasing temperature.

To measure the flow density and temperature of materials, a FLIR thermal imager with sensitive microbolometers elements is used, which register the radiation of the melt at a wavelength of 8 microns. The functionality of the FLIR thermal imager allows you to study the processes of heating, melting of materials in the focal area of the BSF. Most likely, the processes of heating, melting of the material in the concentrated solar irradiation (CSI) flow are accompanied by the processes of radiation, convection, as well as losses due to thermal conductivity, depending on the phase state, the granularity of the material in the initial state. Such features should be taken into account when modeling the process of interaction of concentrated solar radiation with materials.

4. Equipment and stands for melting, synthesis, heat treatment and complex testing of materials and various technical products

BSF is a complex optical-mechanical, electrical system and its successful operation requires appropriate instrumentation, instruments and specific equipment for high-temperature processes. These include, for example, meters -radiometers, calorimeters, digitization systems (system of technical view-STV), photometers, actinometers, computer interface systems, pyrometers, thermal sights etc., modern theodolites, levelers, narrow-aperture lasers, spotting tubes and others.

In the focal zone of the BSF various melting units are installed - furnaces of "bucket" and "rotary" types, for example. **Figure 4** shows the scheme of a water - cooled furnace of the plate type – this is a typical scheme of melting furnaces.

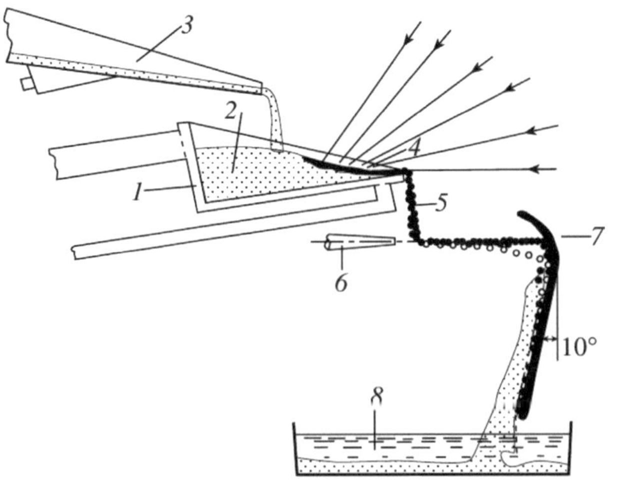

Figure 4.
Water-cooled plate type furnace: (1) metal body, (2) material being smelted, (3) material feed, (4) melt zone, (5) melt stream, (6) water jet, (7) barrier, (8) cold water reservoir.

The ladle furnace is designed for experimental melting in small portions with constant stirring during the melting process in order to achieve homogeneity of the fused material. The furnace allows you to cook the charge, lighten the glass mass and then carry out a one-time production of the entire portion of the melt. The control of the furnace by means of swings makes it possible to mix the liquid phase during the melting process to achieve better clarification and homogenization. The highest melting performance is achieved when the zone with the maximum concentration of light energy is located on the vertical surface of the melted materials. Therefore, for high-performance melting, it is necessary that the irradiated surface of the material forms a right angle with the optical axis of the furnace. However, in the case of powder materials, this is not possible. Therefore, blocks of a certain size are pressed from powder material. A "rotary" type furnace is convenient for melting compressed blocks.

In a "rotary" type furnace (**Figure 5**), the charge is cooked on the surface and during the melting process, the glass mass is clarified in the hot zone of the focal area. Frit is obtained from clarified glass for further ceramic processing. The "rotary" type furnace is quite simple and convenient for operation, and is currently the main BSF equipment for melting high-temperature materials. The ladle furnace is designed for experimental melting in small portions with constant stirring during the melting process in order to achieve homogeneity of the fused material and to avoid delamination of the melting products.

This type allows you to overheat the liquid state and unload the melted mass once. There is a swing system that allows you to form a continuous flow of melt and adjust its speed. The materials of the procedure are melted in a ladle furnace mounted on a coordinate trolley, which manifests itself in the degree of freedom (**Figure 6**).

The upper inner part of the furnace is used as a reflector of energy coming from the lower part of the concentrator. Such a device provides additional heating to the

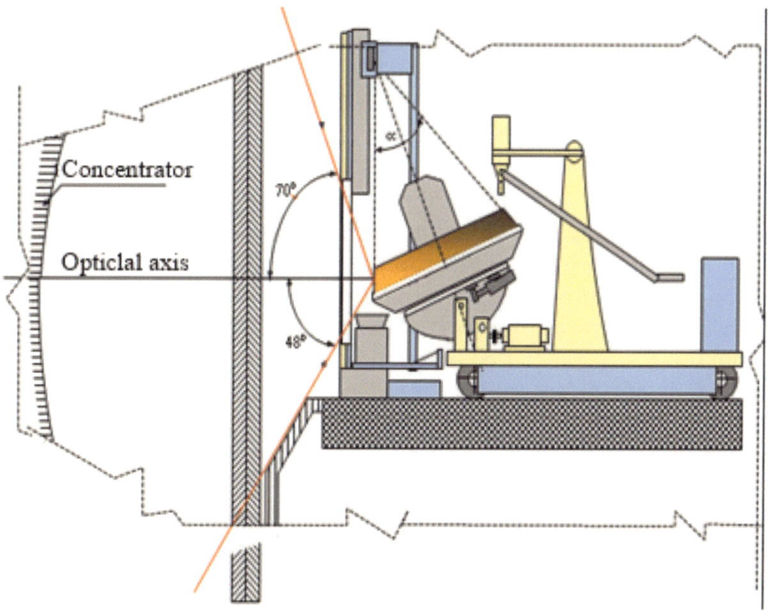

Figure 5.
General view of rotary smelting furnace, located at focus of the BSF.

Figure 6.
A melting furnace of the "bucket" type.

molten material and is maintained in a horizontal position. When the material is pumped in the liquid phase, it moves from one part of the furnace to the opposite through the focal spot. The oven allows you to cook the charge, lighten the glass mass. The control of the furnace by means of swings makes it possible to move the liquid phase during the melting process and achieve better clarification and homogenization. Passing through the focal spot, the material is constantly heated and mixed, thus achieving homogeneity of the fused material. Passing through the focal spot, the material is constantly heated and mixed, thus achieving homogeneity of the fused material.

Along with the advantages of high purity and homogeneity, preservation of stoichiometry, implementation of highly gradient conditions, the possibility of overheating the melt in the air and its subsequent cooling in a controlled mode seems promising. Such conditions make it possible to stabilize new phases, vary the phase composition, morphology of the target material, and as a consequence, its properties.

When the surface of the melted material is inserted as much as possible perpendicular to the flow of concentrated solar radiation, a high melting performance is observed. Consequently, the surface of the material irradiated by the concentrator must have the shape of a vertical wall and form a right angle with the optical axis of the solar furnace. In practice, it is not always possible to obtain a vertical wall due to the powder state of the material. To obtain a vertical wall, the powdered material is pressed into plates. Such materials are easy to melt in water-cooled "rotary" type melting furnaces. To ensure these conditions, a water-cooled "rotary" type furnace has been created.

In a "rotary" type furnace, the charge is cooked on the surface and in the process of draining in the hot zone of the focal spot, the glass mass is clarified. Frit is obtained from clarified glass for further ceramic processing. The "rotary" type furnace is quite simple and convenient to operate, currently it is the main BSF equipment for melting high-temperature materials. When synthesizing glasses, special attention should be paid, firstly, to the degree of overheating of the melt, and secondly, to its cooling rate. It is advisable to use ultrafast quenching methods that allow increasing the melt cooling rate to 10^4 degrees/s.

The introduction of concentrated solar radiation from the concentrator to the target can also be varied with the help of special shutters and a screen made of metallic aluminum and cooled with water. In this case, it is possible to sequentially and/or simultaneously inject a concentrated stream onto the sample. Pyrometric measurements revealed that with gradual input of the flow, the heating rate was 1000 degrees/min, and with rapid input of the flow - 700÷ 1000 degrees/sec.

5. Features of high-temperature processes

In materials science there is a fundamental triad of obtaining materials with desired properties: "method of synthesis - morphology - properties". It should be noted that materials with a favorable combination of different properties can easily be obtained by liquid state quenching. Materials obtained from the melt exhibit high values of mechanical and dielectric properties. Of course, such materials are widely used in various industries. As for melts, it should be noted that, unlike ordinary liquids, crystal-like groups, microcrystallites, are present in the structure of melts. The mutual arrangement of groups in the melt strongly affects the structure and properties of the resulting material. On the other hand, the quality of the melt is determined by the rate of heating of the substance to the melting temperature and above, as well as the rate of cooling of the melt.

The use of solar technologies makes it possible to increase the heating rate hundreds of times and obtain a structure from clusters of a certain composition, using the methods of fast (10^3 degrees/s) and ultrafast (10^4 degrees/c) tempering. Thus, the modeling of the processes of heating and cooling of materials in the flow of CSI is of both scientific and practical interest. Heating. The complete equation of the heating process will be written as

$$\frac{dT_s}{dt} = -\frac{\alpha}{c\rho d}(T_s - T_0) - \frac{\varepsilon\sigma_0}{c\rho d}\left(T_s^4 - T_0^4\right) + \frac{(1-R)}{c\rho d}E \tag{1}$$

where α is the proportionality coefficient, called the heat transfer coefficient, W/ (m^2 K); c is the specific heat capacity W/kgK; ρ is the density g/cm^3; d is the layer thickness, m; T_s is the surface temperature of the body and T_0 is the ambient temperature, K; ε is the degree of blackness, σ_0 - Stefan-Boltzmann constant, E is the density of the flux of concentrated solar radiation in units (W/m^2); R is the reflection coefficient of the heated material. The equation consists of three terms and the first term describes convective heat transfer, the second is related to heat losses due to thermal radiation, and the third is due to the absorption of solar radiation energy.

We can state that the process of heating a material in a field of concentrated solar radiation consists of three parts: heating a solid material to melting; the transition of a solid material into a liquid - melting, heating a liquid material. The boundary temperature values can be determined from the equation:

$$\frac{dT_s}{dt} = 0, \text{at } T_s = T_m \tag{2}$$

T_m is melting temperature.

The incoming heat Q is balanced with the melting heat Q_m, that is

$$Q - Q_m = Q - \lambda m = Q - \lambda \rho S d = 0 \tag{3}$$

$$Q = \left[(1 - R)E - \alpha(T_m - T_0) - \epsilon \sigma_0 \left(T_m^4 - T_0^4\right)\right] S t_m \tag{4}$$

$$t_m = \frac{\lambda \rho d}{\left[(1 - R)T_s - \alpha(T_m - T_0) - \beta\left(T_m^4 - T_0^4\right)\right]} \tag{5}$$

where, λ is the specific heat of melting, J/kg; m is the mass of the material, kg; S is the surface area absorbing solar radiation, m2. The initial conditions for the studied materials (pyroxene rocks) were chosen as follows: c = 711 J/kgK, ρ = 3.2 g/cm^3; α = 100 J/(m^2 K); d = 0.05 m; T0 = 320 K; E = 750 W/cm^2; R = 0.15; Tm = 1660 K; λ = 4200 J/kg.

The calculation was carried out in the MATLAB program. **Figure 7** shows the dependence of the temperature of a material sample on the time of exposure to a concentrated stream of solar radiation.

Figure 8 clearly demonstrates the process of heating the material and shows the non-monotonic nature of the temperature change over time in three sections. At the beginning, heating of the solid material is observed until it passes into the liquid state. The duration of the first section is 80 seconds and has a speed of 1385 deg./s.

It is known that as the temperature of the body rises, the rate of thermal motion and the amplitude of oscillations of its atoms should increase, which are forced to move away from each other over long distances, i.e. distant order disappears. And the destruction of the crystal lattice begins, i.e. the solid melts.

In the second section, melting is observed, which lasts about 100 seconds. At the end, an equilibrium state of the liquid is established. In the third section, the liquid material is heated. As can be seen from the curve, the maximum flux of incident solar radiation corresponds to heating saturation.

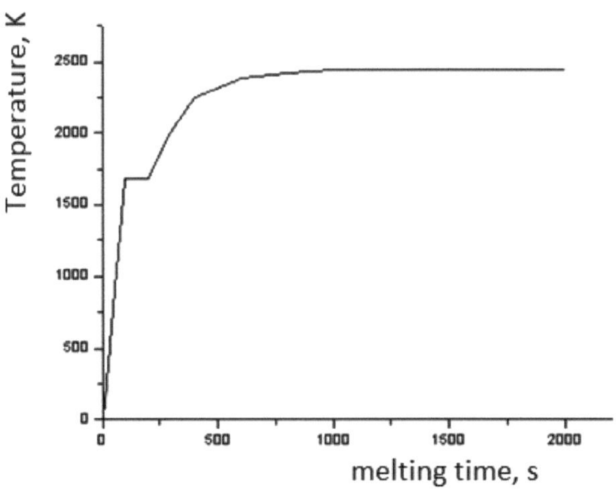

Figure 7.
The dependence of temperature on the time of exposure to a concentrated stream of solar radiation.

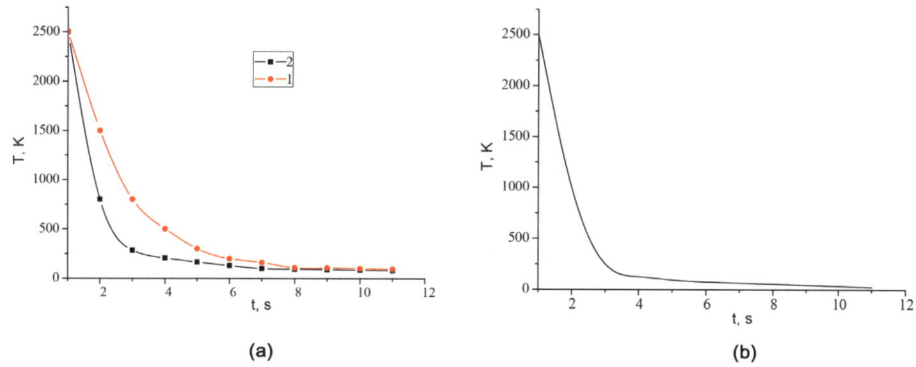

Figure 8.
Melt cooling curves on a solar furnace by the "firecracker" method, with heat transfer coefficients a) 1-α = 500 J/(m² K), 2- α = 1000 J/(m² K). hk = 100mkm, v = 1 m/s. b) Heat transfer coefficient α = 10,000 J/(m² K), hk = 10mkm, v = 10 m/s.

5.1 Cooling

It follows from the above that the cooling rate of the melt is the determining factor in the degree of amorphism of the quenched material. In turn, the cooling rate is determined by the conditions of heat transfer, the temperature of the melt, the material of the hardening system, etc.

There is such a value of the cooling rate Vc, which depends on the level of thermophysical parameters of the material and the nature of the interaction between particles and fluctuates over a wide range (from 10^2 deg./s for inorganic glasses and melts of certain metals to 10^6–10^8 deg./s for metals). High cooling rates are typical for small thicknesses of the cooled melt. In this work, the melts were cooled in three ways: - collapse of the melt according to the "hammer-anvil" principle between water-cooled rods ("clapperboard"), in which the melt is quenched at high speed; – pouring liquid drops into water; – cooling on the water-cooled surface of the substrate.

The amount of heat released through the surface of the body S per unit of time depends on the temperature difference between the surface of the body Ts and the environment T_0 ($T_s > T_0$):

$$\frac{dQ}{dt} = -\alpha(T_s - T_0)S \tag{6}$$

If we assume that the temperature distribution inside the droplet is uniform, as well as the heat input, we obtain

$$\Delta Q = cm\Delta T \tag{7}$$

Eq. (6) can be rewritten as:

$$\frac{dT_s}{dt} = -\frac{\alpha}{cm}(T_s - T_0)S \tag{8}$$

where c and m are the specific heat and the mass of the droplet, respectively.

It should be noted that, as a result of melt compression, its contact surface changes with time; its shape changes from spherical to lamellar. This means that S in (8) is a function of time S = S(t).

Such a dependence can be revealed within the framework of the following boundary conditions. We assume that when compressed between two rods moving against each other with an average speed v, a spherical drop of volume V after time t gradually takes a disk-like shape (for example, due to the wetting effect) of radius R and height h, depending on time. So the volume V is preserved, and the height of the drop changes as h(t) = d-vt. This allows you to determine the area of the end face of the disk S, as follows:

$$S(t) = \frac{\mathrm{V}}{h(t)} = \frac{V}{d - vt} \tag{9}$$

where d is the diameter of the ball.

Taking into account the fact that heat is transferred simultaneously from both ends of the disk, we can replace S with 2S, and from expressions (9) and (8) we can come to

$$\frac{dT_s}{dt} = \frac{\alpha}{cm}(T_s - T_0)\frac{v}{d - vt} \tag{10}$$

To solve, we will need to find the following integral

$$I = \int_0^t \frac{dt}{d - vt} = -\frac{1}{v}\ln\left(\frac{d - vt}{d}\right) \tag{11}$$

Thus, for the general solution of Eq. (10) we obtain

$$T_s = \begin{cases} T_0 + (T_{s0} - T_0)e^{-\frac{2\alpha V}{cmv}\ln\left(\frac{d}{d-vt}\right)} \text{ при } t \le t_0 \\ T_0 + (T_{s1} - T_0)e^{-\frac{2\alpha Vt}{cmh_k}} \text{ при } t > t_0 \end{cases} \tag{12}$$

where

$$T_{s1} = T_0 + (T_{s0} - T_0)e^{-\frac{2\alpha V}{cmv}\ln\left(\frac{d}{h_k}\right)} \tag{13}$$

Where h_k is the final thickness of the disk and

$$t_0 = \frac{d - h_k}{v} \tag{14}$$

The process of cooling by the "damper" method under the condition α = 500 J/(m2K) (curve 1), α = 1000 J/(m^2K) (curve 2) has the character of a nonmonotonic decrease in time (**Figure 8**).

From **Figure 9**, it was revealed that when the melt is cooled by pouring it into water, the cooling rate is about 10^3 K/s. While it follows from **Figure 10** that the cooling of the melt on the surface of the water-cooled substrate proceeds at a rate of 20 K/s at d = 0.1, α = 1000 J/(m^2 K).

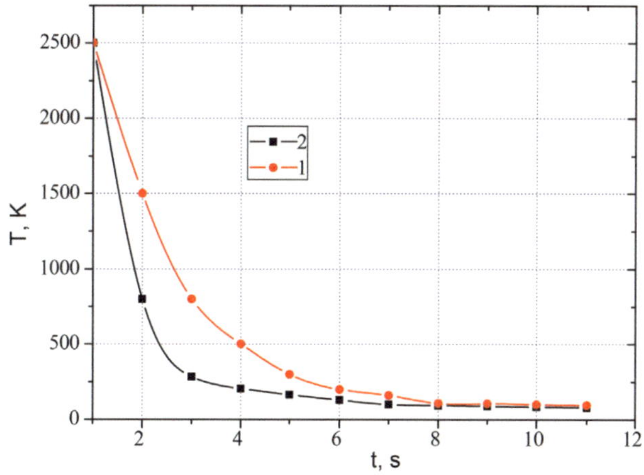

Figure 9.
The curve of melt cooling in a solar furnace by draining liquid droplets into water at α = 1000 W/(m² K).

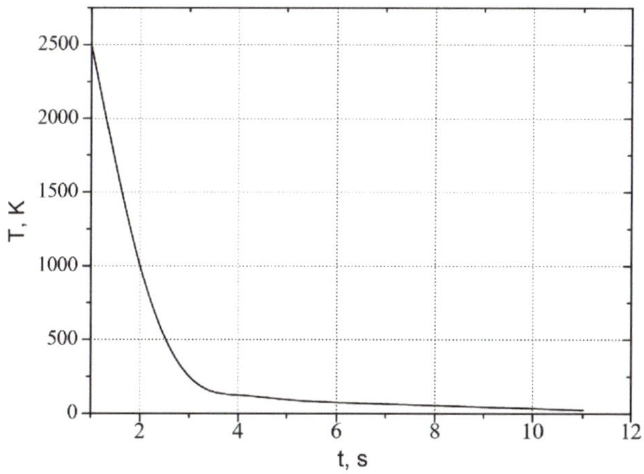

Figure 10.
The cooling curve on the surface of a water-cooled furnace at d = 0.1, α = 1000 J/ (m² K).

The analysis of the curve in **Figure 9** shows that the melt cooling rates in a solar furnace by draining liquid droplets into water are about 10^3 K/s. **Figure 10** shows the cooling curve of the material (melt) in a solar furnace by cooling on the surface of a water-cooled substrate at d = 0.1, $\alpha = 1000$ J/(m^2 K). The analysis of the curve in **Figure 11** shows that the melt cooling rate in a solar furnace by cooling on the surface of a water-cooled furnace is about 20 K/s. Thus, by choosing the melt cooling method, different cooling rates can be achieved: 10^2; 10^3 and 10^4 K/s. For pyroxene melts at high cooling rates $T > 10^3$ K/s, the condition of homogeneous nucleation and growth of crystalline grains is fulfilled. In this case, the grain size is determined by diffusion.

$$d \sim \tau U \sim C\left(\frac{\Delta T^3}{T_m^2}\right)\exp\left(-\frac{E}{kT_{cr}}\right) \tag{15}$$

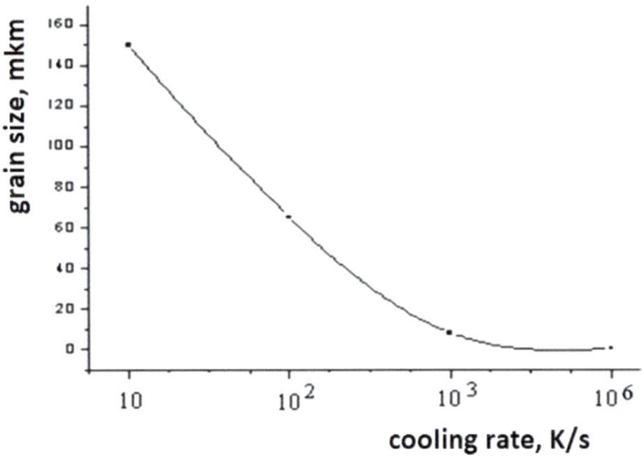

Figure 11.
Dependence of the particle size of the material on the quenching rate.

where τ is the average grain growth time corresponding to the crystallization time, s; U is the grain growth rate of microns/s; C is a value depending on the cooling rate, melting temperature and enthalpy, surface tension, specific volume of the solid phase and Debye frequency of microns; T_{sg} is the melt crystallization temperature, K; DT is the value supercooling ($DT = T_m - T_{cr}$), K; E is the effective activation energy of diffusion, eV.

Figure 11 shows the dependence of the grain size of the material on the quenching rate. It can be seen from **Figure 11** that approximating this dependence to the maximum possible quenching rate makes it possible to determine the size of clusters in the liquid state of matter. To obtain a hardened material with nanosized particles, it is necessary to cool the melt at a rate above 10^6 deg./s.

The processes of heating, melting and cooling of pyroxene rocks in a stream of concentrated high-density solar radiation are well described within the framework of a mathematical model, taking into account the initial conditions. The results of the calculations are in good agreement with the experimentally observed ones.

The strongest influence on the dispersion of the fused material is exerted by the rate of cooling of the melt, which can be carried out by three methods: arbitrarily on the surface of a water-cooled substrate; pouring into water (hardening); anvil - by the collapse of the melt with copper rods (superhardening). Thus, to obtain a fused material with nanosized particles, it is necessary to superharden the melt at a rate above 106 deg./s.

Thus, during the synthesis of materials under the influence of concentrated solar radiation, the melt can be overheated and cooled at different rates, which makes it possible to change and fix the states of a certain phase composition and microstructure, thereby regulating the properties of the resulting material.

Thus, the processes of heating, melting and cooling of materials (pyroxene rocks) on a large solar furnace were simulated.

The proposed mathematical model, taking into account real conditions, quite well describes the processes of heating, melting and cooling of pyroxene rocks in a stream of concentrated solar radiation. It turned out that the dispersion of the obtained material depends on the rate of cooling of the melt, which is set by the method of its implementation. For example, a nanomaterial can be obtained by cooling the melt at a rate above 10^6 deg./s.

Author details

Mukhammad-Sultan Payzullakhanov[1*], Rasul Akbarov[1] and Lola Suvonova[2]

1 Material Sciences Institute, Academy of Sciences Republic of Uzbekistan, Tashkent City, Uzbekistan

2 Tashkent Institute of Irrigation and Agricultural Mechanization Engineers, Tashkent, Uzbekistan

*Address all correspondence to: fayz@bk.ru

References

[1] IRENA. Global Energy Transformation: A Roadmap to 2050. Abu Dhabi: International Renewable Energy Agency; 2018

[2] Fouquet R. Historical energy transitions: Speed, price and system transformation. Energy Research & Social. Science. 2016;**22**:7-12. DOI: 10.1016/j.erss.2016.08.014

[3] United Nations. Pathways to sustainable energy. Accelerating energy transition in the UNECE region. In: ECE Energy Series No. 67. Geneva: United Nations; 2020

[4] Rakhimov YE et al. Analysis of the solar energy potential of the Republic of Uzbekistan. Applied Solar Energy. 2017; **53**(4):344-346

[5] World Energy Transitions Outlook 2023: 1.5°C Pathway; Preview [Internet]. Available from: www.irena.org, https://www.irena.org/Publications/2023/Mar/World-Energy-Transitions-Outlook-2023

[6] Ассоциация "Глобальная энергия" посвятило развитию энергетики нашей страны статью "Амбиции Узбекистана в ВИЭ" [Internet]. [cited 2023 Sep 20]. Available from: minenerg y.uz, https://minenergy.uz/ru/news/view/1540

[7] Avezova NR, Akmal Vokhidov AA, Farmonov ND. Renewable energy: Challenges and solutions. Applied Solar Energy. 2019;**55**(2):149-152. DOI: 10.3103/S0003701X1902004X

[8] Avezova NR, Toshov J, et al. Renewable energy: Scenario and model of development. Applied Solar Energy. 2019;**55**(6): 533-543. DOI: 10.3103/S0003701X 19060021

[9] Подписаны соглашения с компанией "ACWA Power" по строительству двух солнечных фотоэлектростанций в Узбекистане. Также, в Ташкентской области начато строительство солнечной фотоэлектростанции мощностью 400 МВт [Internet]. [cited 2023 Sep 20]. Available from: minenergy.uz, https://minenergy.uz/ru/news/view/2634

[10] Tashkent N. International Solar Energy Institute Renewable Energy Development in Uzbekistan: Current Status, Problems and Solutions [Internet]. Available from: https://www.carecprogram.org/uploads/5.-ISEI_RE-development-in-UZB.pdf

[11] Liu Y, Li F, Ren J, Ren G, Shen H, Liu G. Solar thermal power generation technology research. In: E3S Web of Conferences. Vol. 136. Paris, France: EDP Sciences; 2019. p. 02016. DOI: 10.1051/e3sconf/201913602016

[12] Daniel Fernández-González I, Ruiz-Bustinza CG-G, Noval JP, Mochón-Castaños J, Sancho-Gorostiaga J, Verdeja LF. Concentrated solar energy applications in materials science and metallurgy. Solar Energy. 2018;**170**: 520-540. DOI: 10.1016/j.solener. 2018.05.065

[13] Flamant G, Ferriere A, Laplaze D, Monty D. Solar processing of materials: Oportunities and new frontiers. Solar Energy. 1999;**66**(2):117-132

[14] Charpentier L, Dawi K, Eck J, Pierrat B, Sans J, Balat-Pichelin M. Concentrated solar energy to study high temperature materials for space and energy. Journal of Solar

Energy Engergy. 2011;**133**(3): 031005-031013

[15] Herranz G, Rodríguez GP. Uses of concentrated solar energy in materials science. In: Rugescu RD, editor. Solar Energy. InTech; 2010. Available from: http://www.intechopen.com/books/solar-energy/uses-of-concentrated-solar-energy-in-materials-science. ISBN: 978-953-307-052-0

[16] Abdurakhmanov AA, Faiziev SA, Akbarov RY, Suleimanov SK, Rumi MK, Paizullakhanov MS, et al. Properties of pyroxene glass ceramics, heat treated in the big solar furnace. Applied Solar Energy. 2009;**45**(1): 45-47

[17] Faiziev SA, Paizullakhanov MS, Nodirmatov EZ, Akbarov RY, Zufarov MA. Synthesis of pyroxene pyroceramics in large solar furnace with ZrO_2 crystallization nucleator. Applied Solar Energy. 2008;**44**(2): 139-141

[18] Adylov GT, Akbarov RY, Singh D, et al. Crystallization of μ-and α-cordierite in glass obtained via melting by concentrated radiant flux. Applied Solar Energy. 2008;**44**:135-138

[19] Kh Suleimanov S, Dudko OA, Kulagina NA, Ubaidullaev BF. Modification of aluminum alloys in a solar furnace. Applied Solar Energy. 2007;**43**(2):113-114. DOI: 10.3103/S0003701X07020156

[20] Khaitmukhamedov AE. Development dynamics of concentrating solar power technologies. Applied Solar Energy. 2022;**58**:318-321

[21] Kiseleva SV, Kolomiets YG, Popel OS. Assessment of solar energy resources in Central Asia. Applied Solar Energy. 2015;**51**(3):214-218. DOI: 10.3103/S0003701X15030056

[22] Chub VE. Climate Change and Its Influence on Hydrometeorological Processes, Agroclimatic and Water Resources of the Republic of Uzbekistan. Tashkent: Uzhydromet, Nigmi, Voris-Nashiryot; 2007

[23] Trieb F. Global potential of concentrating solar power. In: Conference Proceedings. Solar PACES, 15-18 Sep 2009. Berlin; 2009

[24] Islam M, Huda N, Abdullah AB, Saidur R. A comprehensive review of state-of-the-art concentrating solar power (CSP) technologies: Current status and research trends. Renewable and Sustainable Energy Reviews. 2018; **91**:987-1018. DOI: 10.1016/j.rser.2018.04.097

[25] Maria Simona Răboacă. Gheorghe Badea, Adrian Enache and etc. concentrating solar power technologies. Energies. 2019;**12**(6):1048. DOI: 10.3390/en12061048

[26] Ruidong W, Jun MA. Status and future development prospects of CSP. IOP Conference Series: Earth and Environmental Science. 2021;**687**: 012088. DOI: 10.1088/1755-1315/687/1/012088

[27] Grilixes VA. Solar Space Power Plants. L.: Nauka; 1986. p. 182

[28] Yu AR, Paizullakhanov MS. Characteristic features of the energy modes of a large solar furnace with a capacity of 1000 kW. Applied Solar Energy. 2018;**54**(2):99-109

[29] Guillot E, Rodriguez R, Boullet N, Sans J-L. Some details about the third rejuvenation of the 1000 kWth solar furnace in Odeillo: Extreme performance

heliostats. Conference Paper. Solar PACES 2017. Santiago, Chile. In: AIP Conference Proceedings. 2018. DOI: 10.1063/1.5067052

[30] Riveros-Rosas D et al. Optical design of a high radiative flux solar furnace for Mexico. Solar Energy. 2010;**84**:792-800

[31] Flamant G, Ferriere A, Laplaze D, Monty C. Solar processing of materials: Opportunities and new Frontiers. Solar Energy. 1999;**66**(2):117-132

Chapter 5

Addition of Organic Compounds in Gelatin-Biopolymer Gel Electrolyte for Enhanced Dye-Sensitized Solar Cells

Sundaramurthy Devikala
and Johnson Maryleedarani Abisharani

Abstract

This chapter introduced a new series of organic compound additives like thiophene 2,5-dicarboxylic acid (TDA), sulfanilamide (SAA), 2,6-diamino pyridine (DAP), dibenzo-18-crown-6 (DBC) and 2,6-pyridine dicarboxylic acid (PDA) with gelatin/KI/I_2 consist gel polymer electrolytes for dye-sensitized solar cells (DSSCs) application. Nowadays, it is focusing on biopolymers for preparing gel electrolytes for DSSCs application which is a conventional renewable energy source. Biopolymers are abundant in nature, and they are non-toxic, thermally stable, environmentally friendly, low-cost, and have good mechanical and physical properties. The introduced novel gelatin (GLN) biopolymer-based gel electrolytes play a role in improving ionic conductivity and stability, and it also play a better ability for ionic mobility. The low-cost and commercialized organic additive molecules with electron donors like S, O and N elements were strongly coordinated on the surface TiO_2 and fermi level shift into negative potentials. The organic additive compound SAA achieved a very active additive and easily reduced the recombination reaction between the surface of TiO_2 and I_3^- ions. This phenomenon readily improves the stability and overall η of the DSSC. During the DSSCs process, intrinsic charge carrier transfer between both electrodes as well as the continuous regeneration of the dye molecules. The surface study and conductivity of prepared gelatin-based gel electrolyte with N, S and O-based additives were characterized by fourier transform infrared spectroscopy (FTIR), UV-visible, X ray diifraction (XRD), Electrochemical Impedance Spectroscopy (EIS) and dye-sensitized solar cells (DSC), respectively. Furthermore, to examine the adsorption behaviour of organic additives on TiO_2 (101) surface and negative fermi level shift on TiO_2 surface were analysed by density functional theory (DFT) theoretical study.

Keywords: DBA, TDA, DSSCs, TiO_2, biopolymers

1. Introduction

DSSC has a conventional renewable energy source and ecologically interest owing to their simple design, easy fabrication, low production cost and high power conversion effiency (PCE) [1, 2]. O'Regan and Grätzel were the first to achieve PCE of 7.1% for DSSCs devices in 1991. Currently, DSSCs achieved a PCE 15.1% efficiency increase by using two metal-free organic dyes [3, 4]. In DSSCs device, there are three major constituents as listed below: (1) FTO plate coated using TiO_2 nanocrystal, it will act as a photo anode as well as grip for dye molecules (sensitizers), (2) redox couple I^-/I_3^- act as a charge transfer which contains electrolyte act as an oxidized dye molecule it regenerates continuously and (3) FTO plate act as a counter electrode which is regenerate the electrolyte using platinum coated. Photosensitizers yielding superior photocurrents are crucial for copper-electrolyte-based, highly efficient dye-sensitized solar cells (DSCs). One of the essential constituents in the DSSCs is the electrolyte, which has five main constitutes: (i) glass conductive substrate, (ii) sensitizing dye, (iii) an electrolyte containing a redox couple, (iv) a semiconductor, and (v) a platinized counter electrode. DSSCs have a combination of the combination which affects the performance dramatically [5, 6]. DSSCs failed to propose the liquid electrolytes there are numerous reports based on polymer electrolytes. During the DSSCs process, intrinsic charge carrier transfer between both electrodes as well as the continuous regeneration of the dye molecules. One of the important factors that affect the photovoltaic performance of DSSCs is a solvent type and immersion time for the photoanode. To prepare the solvent chosen for the photo anode affects the adsorption of the dye on the surface of a semiconductor. This liquid electrolyte has some disadvantages, like liquid volatilization, leakage and sealing complications, and voltage fluctuation which have a significant impact on the long-term stability, maintenance and performance of the DSSCs [7–10]. The polymeric chain diffusion of ions by intermolecular interaction would alter the ionic conductivity of the polymer. The blending technique helps to reduce the enhancement of the dielectric constant, the energy barrier of charge carrier ions, subdues melting point, boiling point and viscosity. The advantages of blended gel polymer electrolytes such as better charge carrier mobility, tensile strength, high thermal and chemical stability, flexibility, good fluidity, less spillage, excellent dielectric constant, increasing ionic conductivity and ion transfer rate. The gel-based polymer electrolytes enhanced the long-term stability in DSSCs [11, 12]. Several researchers have prepared gel electrolytes have been used. Polymers such as poly (acrylonitrile), poly (vinyl chloride), poly (acrylonitrile), poly (propylene oxide), poly (ethylene oxide), poly (vinylidene fluoride hexafluoropropylene), poly (vinylidene fluoride), poly (methyl methacrylate), etc., used in DSSCs applications [13–18]. Biopolymers are abundant in nature, and they are non-toxic, thermally stable, environmentally friendly, low-cost and include good mechanical and physical properties [19].

Researchers are currently focusing on biopolymers for prepared gel electrolytes to DSSCs application, such as *chitosan*, cellulose, alginate, lignin, gelatin, agarose, hydroxypropyl cellulose, starch, carrageenan, natrosol, cyanoethylated cellulose, and polymers [20–23]. Chitosan, heparin, soluble starch, cellulose, gelatin, PCA, PVP and so on can be used to synthesizing different nanoparticles that can replace various toxic reagents. Gelatin is a biopolymer, and it has been commonly used in many more applications like electrochemical devices, biomedical applications, packing applications, water packing, food industries and tissue engineering [24–27]. Currently, researchers are introducing innovative inorganic metal oxide and organic compound for additives into the gel or liquid electrolytes to improve the PCE and stability of DSSCs. While in the DSSC framework, is considered as the major component, a number of metal oxide geometries are being initiated such as copper oxide (CuO), nickel oxide (NiO), titanium dioxide (TiO_2), zinc oxide (ZnO). The starting material for mixed metal oxide (MMOs) using layered double hydroxides (LDHs) has demonstrated superior performance for DSSC application due to their relatively low-cost and forthright preparation process [28–33]. The electrolyte added in the additive consumes a significant effect on the DSSCS of the photovoltaic performance by the transfer of the photoinjected electron at TiO_2 surface to the redox couple electrolyte and reducing the back electron [34]. The oxygen, nitrogen and sulphur-containing (O, N and S) heterocyclic compounds like N-methyl benzimidazole (NMBI), 4-tert-butyl pyridine (TBP), N-alkyl benzimidazole (NABI), benzothiazole thiophene, propyl iso nicotinate, isopropyl isonicotinate, etc., were mostly used for DSSCs. The addition of triphenylamine, carbazole and triphenylamine to the bridged system improves light-harvesting properties and reduces the band gap. The produced dyes exhibit the electrochemical, desired optical and photovoltaic capabilities in addition to being thermally stable [35–38]. The unoccupied gaps between the connected compounds at TiO_2 surfaces enable the electrolyte solution to enter the dye component and trigger recombination before reaching the TiO_2 interface due to their large size, so they tend to leave. It will affect the photovoltaic performance during DSSCs system operation. The added additive plays well to reduce the electron recombination process between the photoanode (TiO_2) and electrolytes and improves the J_{sc}, V_{oc} and PCE of the DSSCs. The electron-rich donor like N and S-based additives are well adsorbed on the TiO_2 surface and thus CB (conduction band) edges rapidly shift into the negative potential to improve the electron lifetime (τ_n) in the DSSCs device [39, 40]. Gelatin, iodine and potassium iodide were mixed up in DMF solvent and kept in magnetic stirrer for few hours to get the gel mixture. The same procedure was followed for all other organic additives. Good stability and high performance of DSSCs device due to incorporation of inexpensive organic additives with rich lone pair O, N and S groups into gel electrolytes which compared to with and without organic additive gel electrolyte.

2. Experimental section

2.1 Preparation of GLN gel electrolytes

Firstly, gelatin 0.30 g was dissolved in glycerol solvent. Then iodine 0.010 g and potassium iodide 0.030 g were mixed in DMF solvent and the mixture was kept in magnetic stirring. The whole mixture electrolyte was stirred continuously at 3 hours for 70°C. After 3 hours stirred, we get gel mixture (dark brown colour).

The glycerol solvent was extra added until to the gelatin polymer 0.30 g dissolved and then additionally added iodine 0.010 g, potassium iodide 0.030 g, SAA 0.010 g in DMF solvent. Afterwards, the whole combination was magnetic stirred well at 3 hours for 70°C to get gel formation. The same procedure was followed for all another organic additive like PDA, TDA, DBC and DAP other gel electrolytes. The selected commercial organic additives structure and photographs of the six gel electrolytes given in **Figures 1** and **2**.

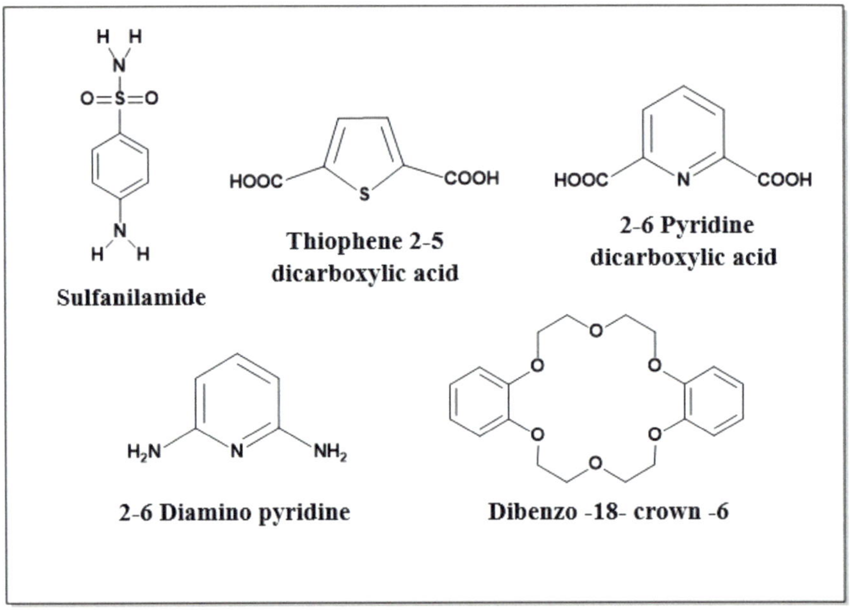

Figure 1.
Chemical name and structure of selected organic additives.

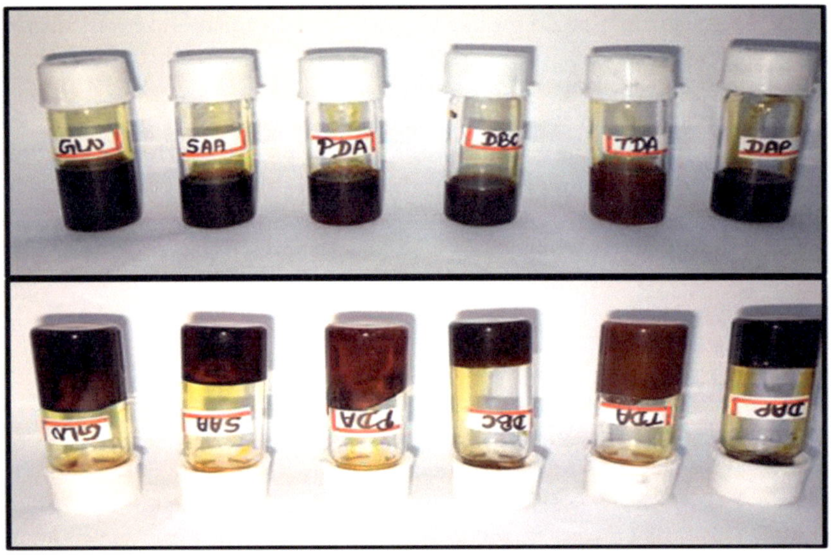

Figure 2.
Images of gel electrolytes without and with additives.

3. Results and discussion

The XRD result of the gel electrolytes was shown in **Figure 3**. Previous literature, the gelatin polymer exhibited broadening peak at 2θ range, 22.35° [41]. From the XRD pattern showed broad peaks around 22° was observed for the entire gel polymer by the amorphous nature of the gel electrolytes which further proved the high

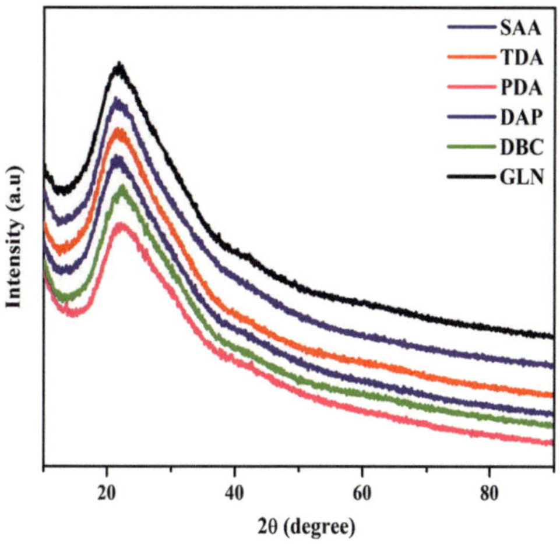

Figure 3.
XRD pattern of gel electrolytes without and with additives.

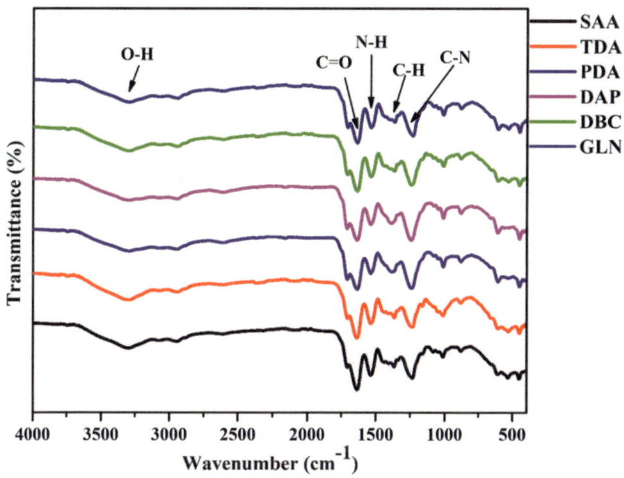

Figure 4.
FTIR spectroscopy of gel electrolytes without and with additives.

conducting electrolyte behaviour. Further it confirms that the organic additives, co-ordinates well with the gelatin polymers matrix and redox couple of gel polymer sample resourcefully. The amorphous phase of such gelatin gel polymer samples is notable parameter in expressing the conductivity nature. Hence, it shows that the conductivity increased with the amorphous nature.

The gelatin gel electrolytes with/without organic additives added upon were measured by FTIR with 400 to 4000 cm^{-1} range. The corresponding FTIR spectroscopy specified in **Figure 4**. From the FTIR, clearly exposes the vibration and stretching frequency of all the gelatin gel electrolyte with/without additives. The broad peak was observed at 3500–3200 cm^{-1} assigned for biopolymer gelatin hydroxyl group. Due to C = O vibrational, stretching frequency intensity peak was attained highly at 1640 cm^{-1}. N-H bending allocated at 1530 cm^{-1} is sharp peak. Small intensity peak was attributed at 1445–1340 cm^{-1} owing to the bending vibration of aliphatic C-H group and noted C-N stretching frequency showed at 1235 cm^{-1} [42]. FTIR clearly shows that the inclusion of organic additives in gelatin gel electrolyte with composition of KI/I_2 changes a small intensity shift, which shows organic additives and redox pair well connection with gelatin polymer.

The gelatin gel electrolyte absorbance range was investigated by UV-visible spectroscopy. **Figure 5** describes the GLN gel electrolyte without additive absorbance range was shown at 313 nm. The gelatin gel electrolytes with organic additives SAA, TDA, PDA, DAP and DBC correspond to 293, 290, 296, 292 and 295 nm, respectively. This small changes in UV-visible absorption may be due to the dissimilar functional groups present in the organic additives interacting with gelatin gel electrolytes. Also, the UV-visible study confirmed the addition of organic additives well incorporated with the I_3^- ion in the gelatin polymer electrolyte. It exhibited a peak at 320-370 nm for all the organic additive-based gel electrolytes. This phenomenon increases the I_3^- ion mobility in the gelatin polymer electrolyte and enhanced well the DSSCs performances.

The morphology structure studied about with and without organic additive in the gelatin gel electrolytes. Without organic additive GLN gel electrolyte given in **Figure 6a**, which showed the surface morphology like fibre nature and without space formation. Gelatin gel electrolyte with organic additives SAA (**Figure 6b**), TDA (**Figure 6c**), PDA (**Figure 6d**), DAP (**Figure 6e**) and DBC (**Figure 6f**) displayed surface morphology like fibre and with space formation which confirms the addition of additives changed the surface of the gelatin gel electrolyte. From the scanning electron microscopy (SEM), it clearly observed that gelatin and additive gel electrolytes increase the amorphous nature and improve the ionic conductivity in DSSCs.

DSC study for gelatin gel electrolytes with and without additives was caried under an annealing temperature of 50 to 300°C with nitrogen atmosphere at 10 k/min. The **Figure 7** displayed decomposition peak for without and with organic additive gel electrolyte. GLN peak revealed at 230°C, which is without additive gel electrolyte.

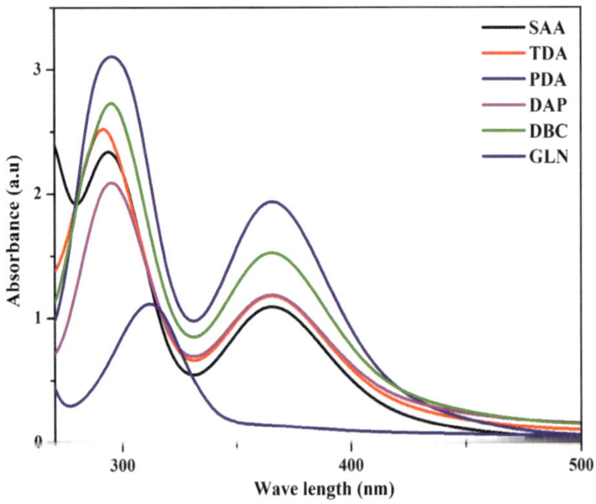

Figure 5.
UV-visible spectroscopy of gelatin gel electrolytes without and with additives.

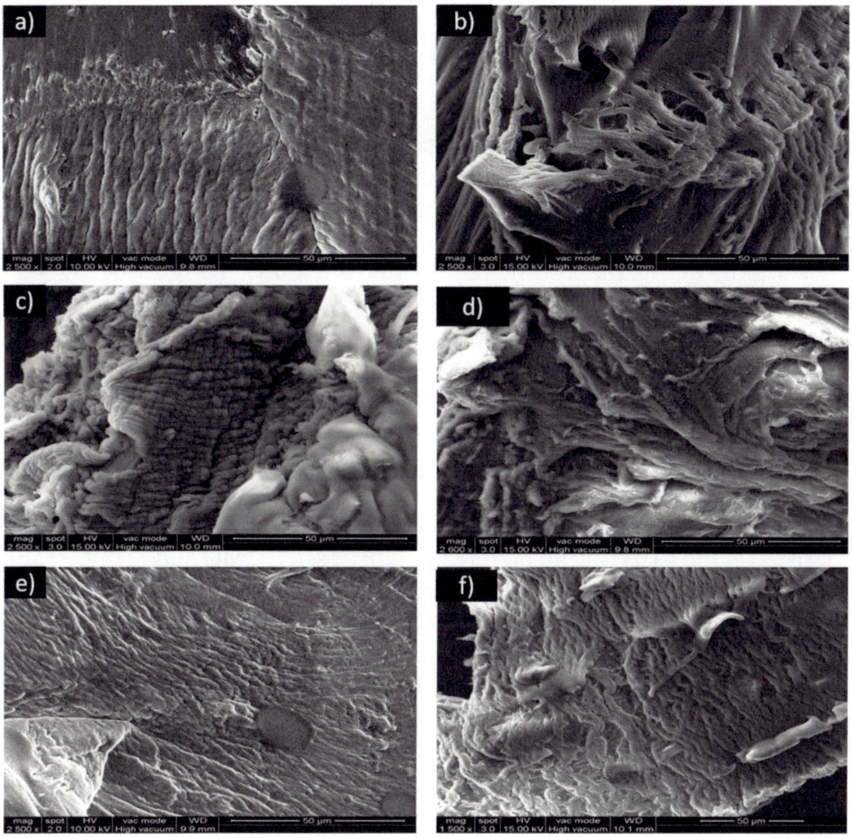

Figure 6.
SEM images of gelatin gel electrolytes (a) GLN, (b) SAA, (c) TDA, (d) PDA, (e) DAP and (f) DBC.

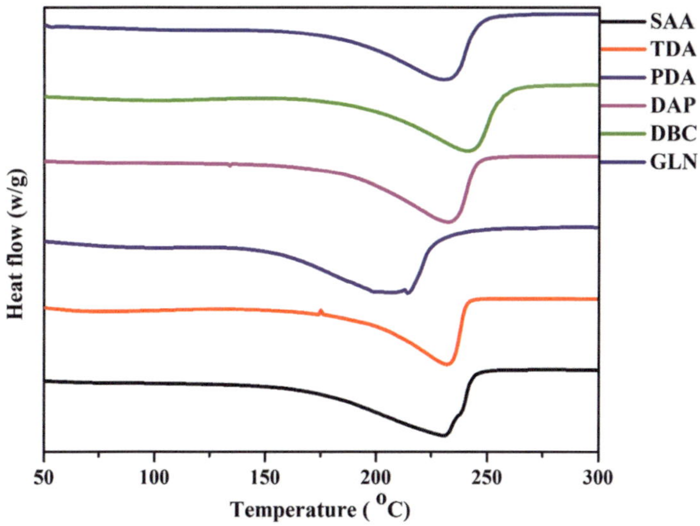

Figure 7.
DSC curves of gelatin gel electrolytes without and with additives.

The organic additives-based gel electrolyte exhibited decomposition peak at 231°C (SAA), 233°C (TDA), 208°C (PDA), 234°C (DAP) and 241°C (DBC), respectively. The variation of the decomposition may be due to the different organic additives and KI/I_2 incorporated into the gelatin gel electrolytes. The broadening decomposition peak confirms the nature of the amorphous gel electrolytes owing to interfacial interaction between the organic additives and polymer. These thermal results prove the gelatin gel electrolytes have a good stability in DSSC performances.

All the organic additives were optimized at a gas phase by Gussian16, and optimized geometries of organic additives compound are shown in **Figure 8**. Highest occupied molecular orbital (HOMO)-lowest unoccupied molecular orbital (LUMO) orbital calculated for the organic additives done by frontier molecular orbital (FMO) analysis optimization at gas phase. **Figure 9** shows the HOMO-LUMO for organic additive compounds. **Figure 9** observes the electronic shift and charge transfer in different orbitals. The transition occurs primarily from the aromatic cloud towards the functional group present in it, which has been noted. The additives containing carboxylic acid and amino functional groups have been mostly noticed. In DBC additive, it has the least efficiency due to the absence of these functional groups, and the orbitals are only localized at the π ring due to no transition between HOMO and LUMO. **Table 1** shows the Energy Gap (Eg) for all the organic additives between HOMO and LUMO.

E_g from the table shown that HOMO-LUMO gap is a high for DBC additive. The FMOs are mainly referred to as the lowest unoccupied molecular orbital (LUMO) highest occupied molecular orbital (HOMO). The photosensitizer for FMOs should be aligned appropriately with the other main two components in the DSSC, i.e., the electrolyte shuttle and semiconductor. The conduction band (CB) of the semiconductors is less stable (destabilized) with respect to LUMO. The order of E_g for all the organic additives as follows DBC > SAA > DAP > PDA > TDA. In the DSSCs process, the entire organic additive HOMO-LUMO gap has a great impact on the transferred electron [28]. The charge transfer process will be more difficult, if the molecule has a higher E_g. In this case, the molecule with the greatest E_g (i.e., DBC), when compared to others has the least efficiency. As a result, it is evident that the E_g plays a very important role in the PCE of DSSCs in the additive molecule.

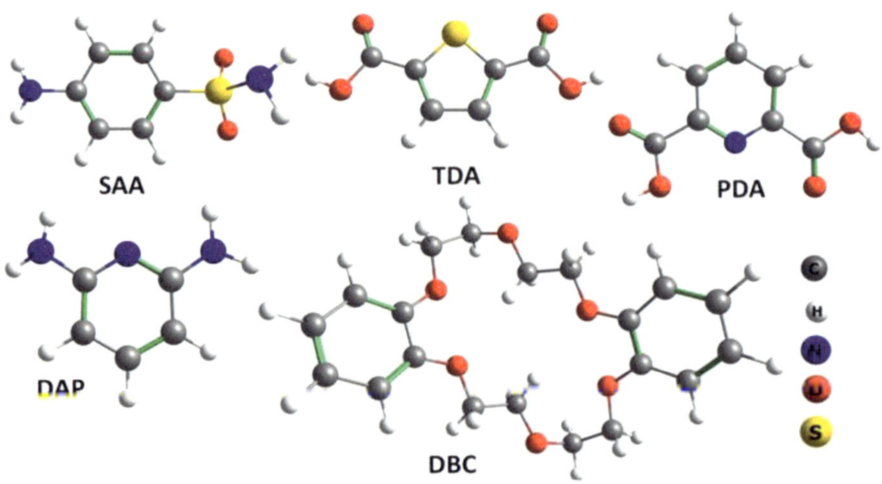

Figure 8.
*Optimized geometries of additive compounds at PBE/6–311++G**.*

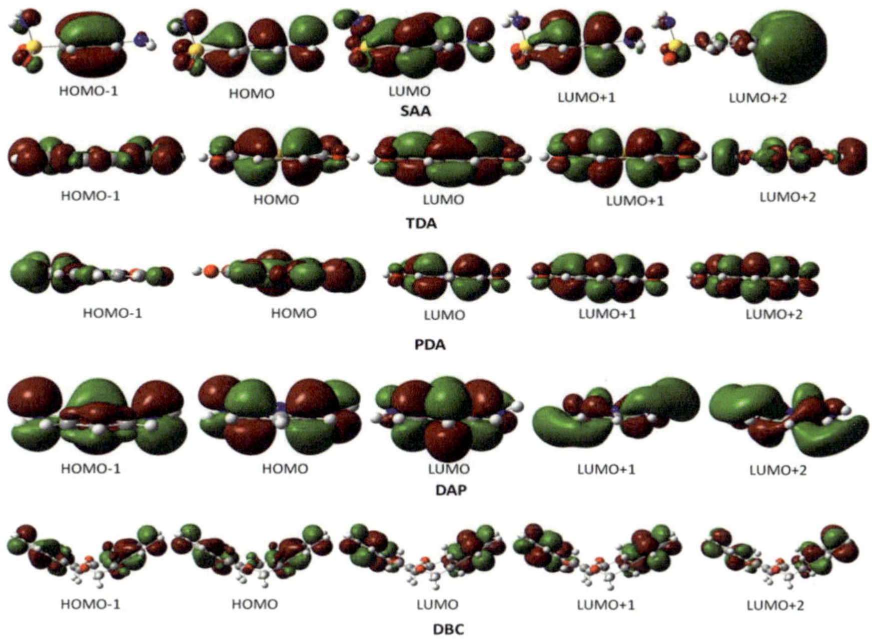

Figure 9.
HOMO and LUMO orbitals calculated for different organic additives in gas phase.

Sample code	E_g in gas phase (eV)	Shortest interaction distances (Additive·TiO_2 Surface)		E_{ads} (kcal/mol)	Fermi energy (eV)
		Covalent bond (Å)	Non-covalent (Å)		
SAA	3.82	2.08 (Ti-O) 2.27 (Ti-N)	1.67 (N-H···O)[c]	−105.88	−3.82
TDA	3.31	1.85 – 1.93 (Ti-O)[a]	1.65 -1.95(O-H$^+$···O)[c]	−368.53	−3.88
PDA	3.32	1.89 – 2.24 (Ti-O)[b]	1.70 (O-H$^+$···O)[c]	−202.35	−3.89
DAP	3.63	2.05 – 2.26 (Ti-N)	1.55 – 1.76 (N-H···O)[c]	−225.62	−3.60
DBC	4.07	—	2.15 – 2.98 (vdWs)[d]	−69.74	−3.16

(a), (b) represents the single and double proton transfer from carboxylate group of additives into TiO_2 surface, respectively. (c), (d) represents the van der Waals interaction distances and hydrogen bonding, respectively.

Table 1.
The interaction distance, E_{ads} and Fermi energy, E_g in gas phase of all TiO_2 surface on organic additives.

The gas phase optimization for additives optimized on the TiO_2 (101) surface. All of these total phase DFT calculations was performed using the CP2K code. During the optimization, the upper two layers were free and lowest two layers was kept fixed surface. After the optimization, the interaction distances, fermi energies, E_{ads} of the additives were calculated and the values given in **Table 1**. The interaction distance between the organic additives and anatase TiO_2 surface have strong non-covalent bonds and covalent bond (i.e., H-bond and vdWs) interactions. The calculated fermi energy and interaction distance are displayed in **Table 1**. Closer analysis of the distance reveals more efficient combinations of strong Ti-O and Ti-N bonds with H-bonded (~ 1.8 to 2.3 Å) interaction on the surface. The strong interaction between the additives and TiO_2 surface is important and significant in the efficiency of the complexes. The shortest H-bond distance and strong covalent bond (1.65 and 1.67 Å) of additive SAA and TDA compounds showed the highest PCE of 5.8 and 5.2%. Due to the formation of covalent bond interaction in PDA and TDA carboxylates encourage the proton transferred into the TiO_2 surface. This DAP additive may be stabilized by two H-bonds (N-H·····O) and Ti-N bonds between the DAP and TiO_2. It is also exhibited an efficiency of 4.3%, which is lower efficiency than the other (SAA, PDA and TDA) additives remarkable in this study.

Previous literature exposes that nature of the geometries and interaction act as vital role in PCE of the DSSCs device [43, 44]. Different functional groups present in all five organic additives strongly interact with covalent and non-covalent (H-bond and vdWs) on TiO_2 surface and it was clearly revealed the **Figure 9**. The carboxylic acid and amino functional group performance has a major significance in stimulating the electrostatic interaction between the organic additives and surface TiO_2. These dealings have a gigantic effect on the charge transfer and PCE of DSSCs. The strong interaction can be computed from E_{ads} values. The organic additives E_{ads} values −105.9 (SAA), −368.5 (TDA), −225.6 kcal/mol (DAP) and − 202.4 (PDA), respectively. Due to the anchoring bond strongly present in between the functional groups in additive and TiO_2 surface, it is exciting to note that the carboxylic functional group present in the TDA and PDA denote the proton from organic additive to TiO_2 surface. There is a strong H-bond formation between the carboxylate anion and TiO_2 (OH) which indicates the ionic state. Thus, these organic additives are powerfully bound to boost the charge transfer and TiO_2 surface, which directly impacts the PCE of DSSCs. The less DBC E_{ads} (−69.7 kcal/mol) which attributed to weak vdWs interaction in the absence of anchoring groups with TiO_2 surface. DBC additive has a no robust electrostatic interaction like other organic additives, which is obviously from distances and interaction sites (**Figure 9**) when compared to other organic additives this is due to the least efficiency of DBC additive. Furthermore, the energy of fermi is not that much affected by different additive molecules. The low fermi energy (≈ 3.5 eV) of the complexes helps in the easy transfer of the charge in DSSCs. After the DFT studies, it was clearly understood that the strong interaction between the organic additive functional groups and surface TiO_2 has important role to act in charge transfer process and hence in improved PCE of DSSCs.

The entire gel electrolyte without and with additives was studied via EIS (Electrochemical Impedance Spectroscopy) under room temperature by the ionic conductivity of. All the measured gel electrolytes scan range from 300 KHz to 400 MHz (0.800 V) as single sign mode and to set Ewe to E. The ionic conductivity of gel electrolytes were calculated from Nyquist plot. The conductivity of gel electrolytes in the order as follows SAA (2.93×10^{-5}) > TDA (2.85×10^{-5}) > PDA (2.77×10^{-5}) > DAP (2.67×10^{-5}) > DBC (2.38×10^{-5}) > GLN (1.72×10^{-5}) and which is shown in **Figure 10**. The conductivity results exposed that inexpensive O, S and N improves the ionic mobility for DSSCs, which contain organic additives in gel electrolytes. In these additives, gel-based SAA electrolytes proved better conductivity than other additives-gel-based electrolytes owing to high electron-donating properties. It consists of S, O and N atoms composed in a structure of the SAA additives. The interaction between the organic additives and TiO_2 surface was confirmed by DFT studies. The organic-based additives gel electrolytes have a moral diffusion of ions molecules

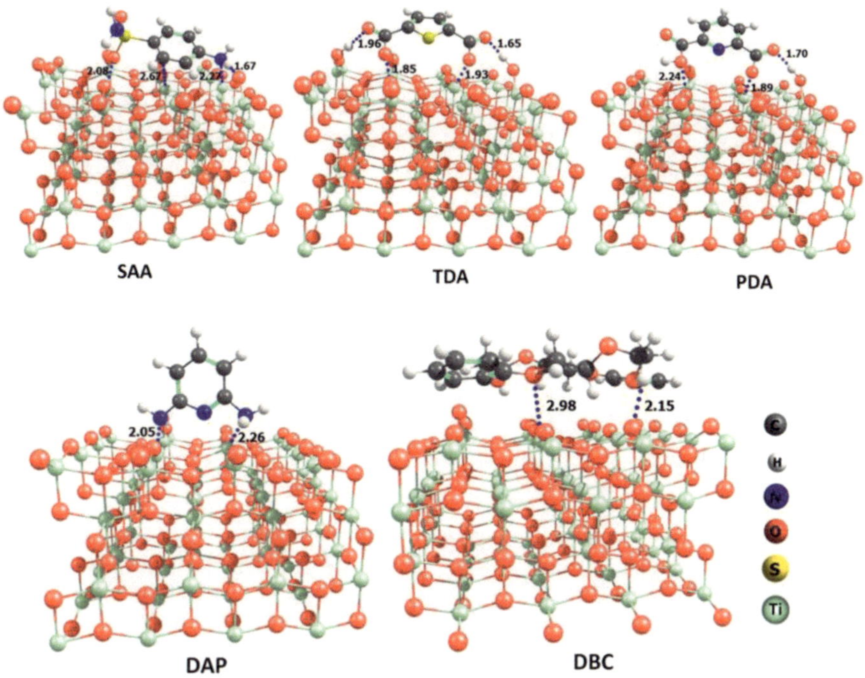

Figure 10.
DFT optimized geometries of organic additives on TiO_2 (101) surface with shortest interacting distances (Å).

owing to the crystalline of gelatin polymer decreases and organic compounds formation of complex of with I_3^- ions prevents the process of recombination. Therefore, by the presence of carboxylic acid and amino linkage groups, highly enhanced ionic conductivity offers a high interaction between the TiO_2 surface and organic additives when compared with GLN gel electrolyte (without additive). The gel electrolyte variation of conductivity is shown in **Figure 11**.

EIS study was examined for photo electrochemical process such as resistance and conductance during DSSCs process. This study mainly discussed about the interfacial charge transfer in DSSCs like charge transfer between the photo anode/electrolyte and platinum/electrolyte reactivity termed as C_μ, R_{ct} and R_{pt} [45]. The EIS data received for devices under darkness was fitted by Z_{fit} software. R_s (R1), R_{pt} (R2), R_{ct} (R3) and C_μ (C3) values collected from fitting equivalent circuit and the values tabulated in **Table 2**. The R_s, R_{pt}, R_{ct} and C_μ corresponds to series resistance, charge transfer between electrolytes and photoelectrodes interface and chemical capacitance of the DSSCs device. **Figures 12** and **13** display chemical capacitance and charge transfer for gel electrolytes. The SAA additive gel electrolyte shows low R_{pt} value of 685 Ω, high R_{ct} of 1674 Ω and C_μ of 5.702×10^{-6} F. From these values compared with other additive-based gel electrolytes due to high because more electron-providing properties of SAA additive.

The without additive GLN gel electrolyte displayed high R_{pt} value of 3623 Ω, low R_{ct} of 411 Ω and C_μ of 4.294×10^{-6} F.

The chemical capacitance and charge transfer resistance results revealed that organic additives are an important key function of complex formation with I_3^- ions. Furthermore, in the fermi level of TiO_2, it act on the TiO_2 surface caused an

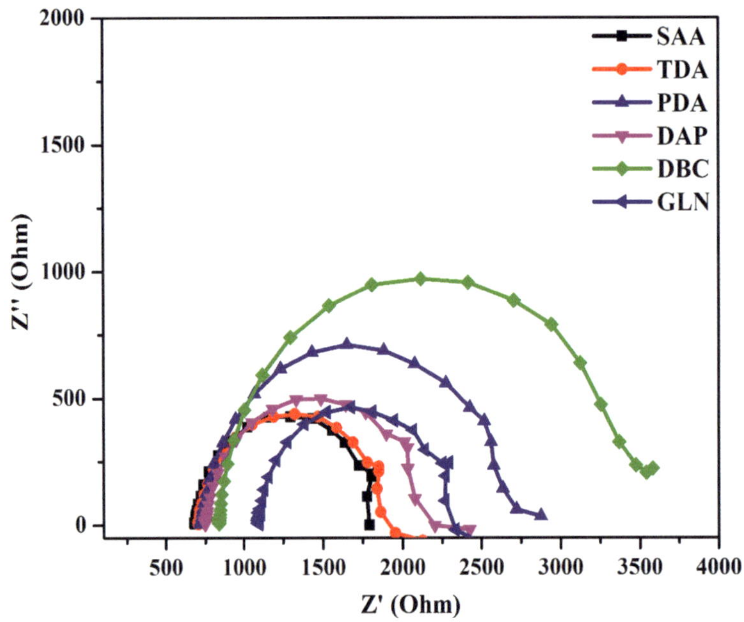

Figure 11.
Nyquist plot of gel electrolytes without and with additives.

Electrolyte system	R1 R_s (Ω)	R2 R_{pt} (Ω)	R3 R_{ct} (Ω)	C3 C_{μ} (F)	σ (S/cm^{-1})	τ_n (ms)
SAA/I^-/I_3^-/gelatin	685	685	1674	5.702×10^{-6}	2.93×10^{-5}	6.6
TDA/I^-/I_3^-/gelatin	705	684	1024	4.926×10^{-6}	2.85×10^{-5}	4.5
PDA/I^-/I_3^-/gelatin	727	727	689	4.861×10^{-6}	2.77×10^{-5}	4.0
DAP/I^-/I_3^-/gelatin	749	762	462	4.842×10^{-6}	2.67×10^{-5}	3.1
DBC/I^-/I_3^-/gelatin	840	1478	446	4.780×10^{-6}	2.38×10^{-5}	3.0
I^-/I_3^-/gelatin(GLN)	1172	3623	411	4.294×10^{-6}	1.72×10^{-5}	2.6

Table 2.
Electron life time and ionic conductivity values of gel electrolytes without and with additives.

immediate change, thereby the charge for the recombination processes get reduced. The augmented chemical capacitance after the addition of additive SAA combined with gel gelatin electrolyte, when the recombination process get decreased for production of good PCE.

Electron life time derived from bode phase plot which was experimental result of EIS measurement and bode phase graph given in **Figure 14**. The recombination kinetics between the photoanode and electrolytes interfaces are relevant to the maximum peak frequency (f_{max}) value from the bode phase plot of gel electrolytes. Electrolytes and electrodes interfaces electron lifetime was calculated by following derivation [46].

$$\tau_n = \frac{1}{2\pi f_{max}} \quad (1)$$

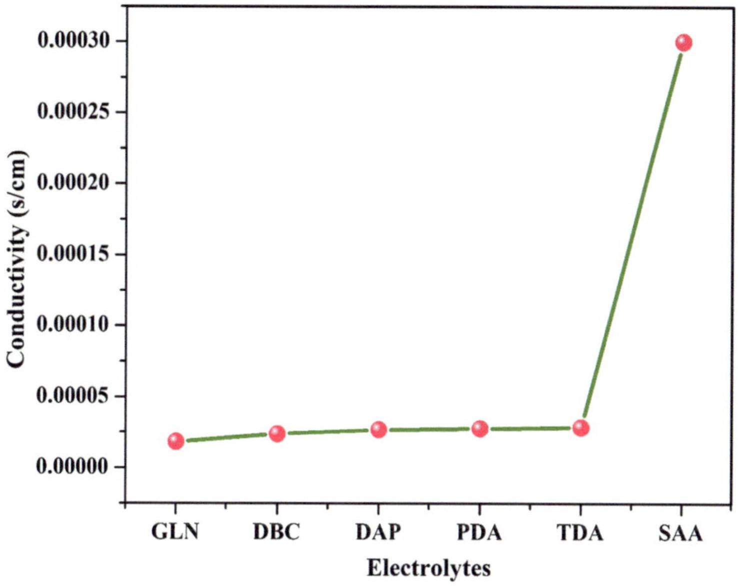

Figure 12.
Conductivity variation of gel electrolytes with and without additives.

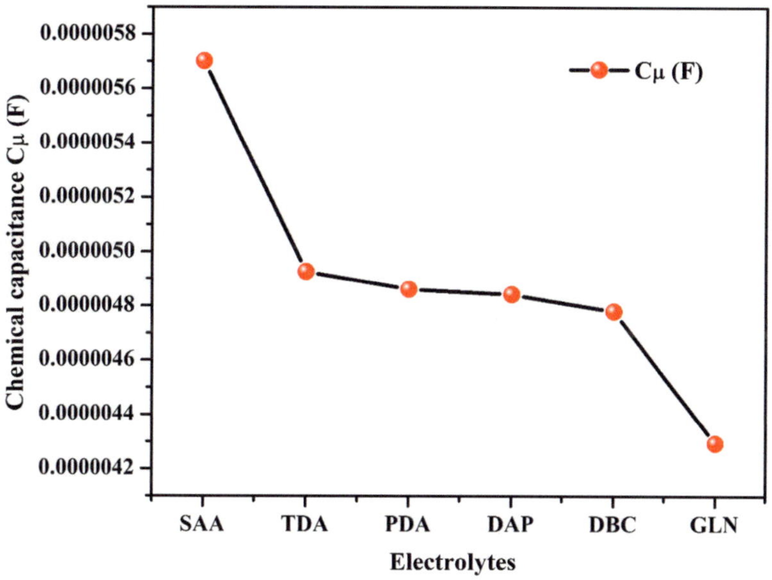

Figure 13.
Chemical capacitance of gel electrolytes without and with additives.

The τ_n values of gel electrolytes given in **Table 2**. The τ_n of gel electrolytes variation of each other owing to presence of various functional groups present in organic additives. The SAA additive incorporated with gel electrolyte showed a high τ_n (6.6 ms) which compared with other organic additives-based gel electrolyte. The without additive GLN gel electrolyte exhibited low τ_n of 2.6 ms. From the τ_n results, increment of τ_n of SAA gel electrolyte showed reduced recombination process among

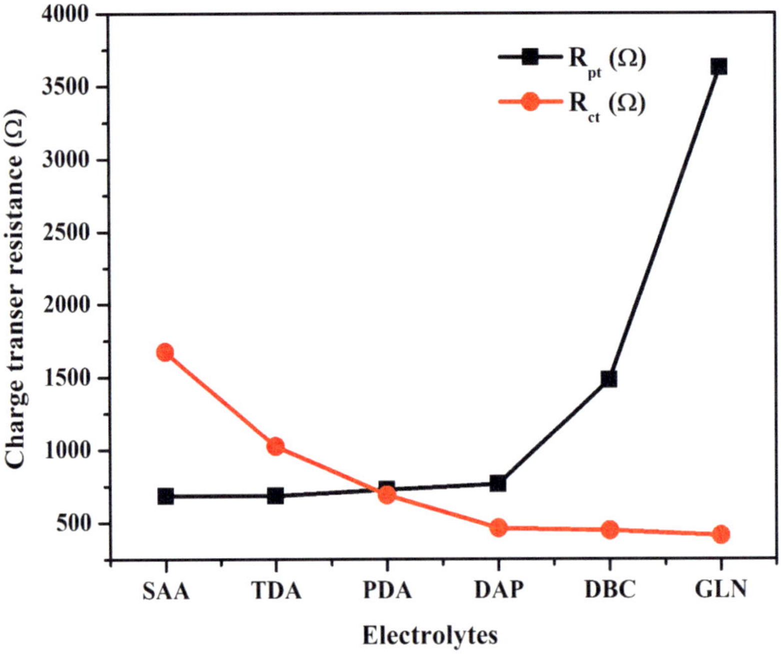

Figure 14.
Charge transfer resistance of gel electrolytes without and with additives.

Electrolyte system	J_{sc} (mA cm^{-2})	V_{oc} (V)	Fill factor (FF)	Efficiency (η)
SAA/ I^-/ I_3^-/gelatin	14.1	790	0.52	5.8
TDA/ I^-/ I_3^-/gelatin	13.0	780	0.51	5.2
PDA/ I^-/ I_3^-/gelatin	12.4	770	0.51	4.9
DAP/ I^-/I_3^-/gelatin	11.3	765	0.51	4.4
DBC/I^-/ I_3^-/gelatin	11.0	760	0.51	4.3
I^-/ I_3^-/gelatin	10.1	752	0.51	3.9

Table 3.
I-V results of gel electrolytes without and with additives for DSSCs device under 1 sun illumination (100 mWcm^2).

TiO_2 surface and I_3^- ions. By modifying the components, we were able to find out with and without the addition of gel electrolyte using DSSCs device, and used for their optical, electrochemical and photovoltaic characteristics [47–49].

I-V measured for gel electrolytes without and with additives was carried by 1.5 Am with sun illumination of 100 mWcm^{-2} and I-V curves are given in **Figure 14**. The photovoltaic results of V_{oc}, FF, J_{sc} and η presented in **Table 3**. The SAA additive gel electrolyte act as a greater electrolyte in DSSCs, which showed the higher PCE of 5.8%. This SAA organic additive has electron-rich donor atoms combined in a structure such as N, O and S, and it stimulates the redox transfer of ions in gel electrolyte via complex formation. Therefore, TiO_2 fermi level has less negative shift evidently proved to be improve the PCE of DSSCs device. Without additive GLN gel electrolyte attained PCE of 3.9% because of GLN gel electrolytes have a less conductivity and charge transfer process. After that, the addition of organic additives into the gel electrolytes improves the PCE DSSCs device. The variation of structure and functional groups in organic additives improves the amorphous nature of gelatin polymer and the good diffusion pathway of I_3^- ions in the gel nature.

The device stability of the gel electrolyte without and with additives was assigned for the stability test at room temperature and its stands for 6 days (**Figure 15**). The calculated efficiency by V_{oc}, FF, J_{sc} and η (**Figure 16**).

4. Conclusion

In this study, good stability and high performance of DSSCs device due to inexpensive incorporation of organic additives with rich lone pair N, O and S groups into gel electrolytes biopolymer gel which compared to without organic additive gel electrolyte. The organic additives (i.e., SAA, PDA, TDA, DBC and DAP) were linked in the surface of TiO_2 that readily fermi level shifts in CB of TiO_2 to a negative potential. The integration of S, O and N-containing organic additives into the gel electrolytes improves the ionic conductivity from 1.72×10^{-5} S/cm^{-1} (GLN gel electrolyte) to 2.93×10^{-5} S/cm^{-1} (SAA gel electrolyte). An increase of the C_μ, R_{ct} and τ_n in DSSCs, additives-based gel electrolyte, clearly explained the decrease in the recombination reaction between the TiO_2/dye/electrolyte interface. Moreover, DFT study suggests that the anchoring groups in organic additives act as a high charge transfer process and their strong covalent bond and non-covalent bond on the TiO_2 surface. SAA additive gel electrolyte attained PCE 5.8% and GLN gel electrolyte reached the PCE 3.9%. This phenomenon proved the importance of N, S and O-containing additives in biopolymer gel electrolyte-based DSSCs performances. Thermal results prove the gelatin gel electrolytes have good stability in DSSC performances. The ionic conductivity is highly enhanced by the presence of carboxylic acid and amino linkage groups that offered a high interaction between the TiO_2 surface and organic additives

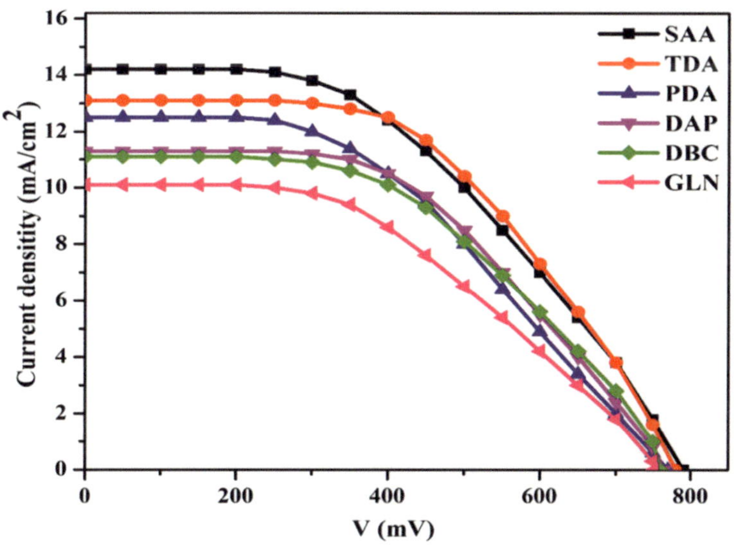

Figure 15.
I-V curve of gel electrolytes without and with additives.

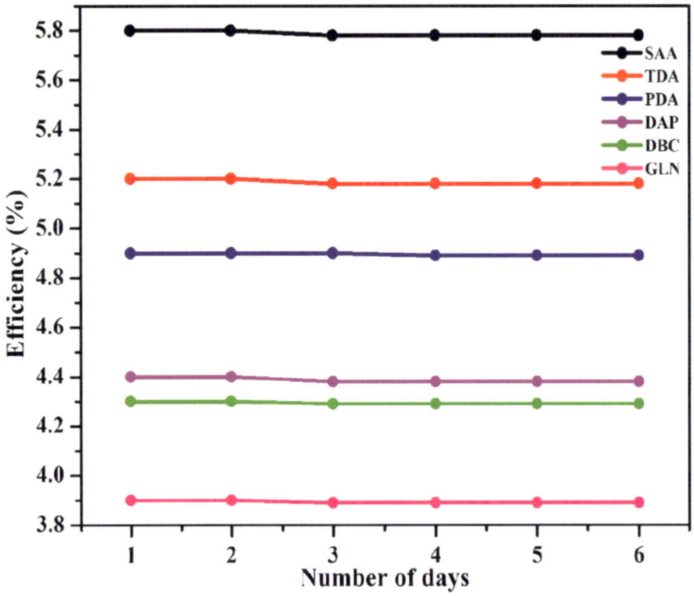

Figure 16.
Stability graph of gel electrolytes without and with additives.

when compared with GLN gel electrolyte (without additive). The structure and functional group variations in organic additives improve the amorphous nature of gelatin polymer and good diffusion pathway in the gel nature. Gelatin with additive gel electrolyte increases the amorphous nature and improves the ionic conductivity in DSSCs. The chemical capacitance is augmented after the addition of SAA additive combined with gelatin gel electrolyte, which decreases the recombination process for the production of good PCE. We introduced inexpensive organic additives with gelatin-biopolymer gel electrolytes, showing high PCE and better replacement of conventional liquid electrolytes in DSSCs applications.

Author details

Sundaramurthy Devikala* and Johnson Maryleedarani Abisharani
Department of Chemistry, College of Engineering and Technology, SRM Institute of Science and Technology, Kattankulathur, Tamil Nadu, India

*Address all correspondence to: lsrinivasan74@gmail.com

References

[1] Chowdhury FI, Buraidah MH, Arof AK, Mellander BE, Noor IM. Impact of tetrabutylammonium, iodide and triiodide ions conductivity in polyacrylonitrile based electrolyte on DSSC performance. Solar Energy. 2020;**196**:379-388

[2] Prajapat K, Dhonde M, Sahu K, Prateek B, Murty VVS, Shirage PM. The evolution of organic materials for efficient dye-sensitized solar cells. 2023;**55**:100586

[3] Peng JD, Wu YT, Yeh MH, Kuo FY, Vittal R, Ho KC. Transparent cobalt selenide/graphene counter electrode for efficient dye-sensitized solar cells with Co2+/3+-based redox couple. ACS Applied Materials & Interfaces. 2020;**12**(40):44597-44607

[4] Sen A, Putra MH, Biswas AK, Behera AK, Gro A. Insight on the choice of sensitizers/dyes for dye sensitized solar cells. 2023;**213**:111087

[5] Masud, Kim HK. Redox shuttle-based electrolytes for dye-sensitized solar cells: Comprehensive guidance, recent progress, and future perspective. ACS Omega. 2023;**8**(7):6139-6163

[6] Grobelny A, Shen Z, Eickemeyer FT, Antariksa NF, Zapotoczny S, Zakeeruddin SM, et al. A molecularly tailored photosensitizer with an efficiency of 13.2% for dye-sensitized solar cells. Advanced Materials. 2023;**35**(5):2207785

[7] Arslan BS, Öztürk N, Gezgin M, Sevindik S, Yılan D, Kumbasar RA, et al. Optimization of fabrication parameters for efficient dye-sensitized solar cells based on dyes containing benzothiadiazole, indoline, and their co-sensitization. Electrochimica Acta. 2023;**184**:142523

[8] Santhosh SG, Balamurugan S. Novel indole-based photosensitizers coupled with PEG-HEC quasi-solid-state electrolyte to improve energy conversion and stability of organic dyes based-dye sensitized solar cells. 2021;**389**:138771

[9] Krishna J, G, Ojha PK, Kar S, Roy K, Leszczynski J. Chemometric modeling of power conversion efficiency of organic dyes in dye sensitized solar cells for the future renewable energy. 2020;**69-70**:10453

[10] Bashir MBA, Rajpar AH, Salih EY, Ahmed EM. Preparation and photovoltaic evaluation of CuO@ Zn(Al)O-mixed metal oxides for dye sensitized solar. Cell Nanomaterials. 2023;**13**(5):802

[11] Alizadeh A, Roudgar-Amoli M, Bonyad-Shekalgourabi S-M, Shariatinia Z, Mahmoudi M, Saadat F. Dye sensitized solar cells go beyond using perovskite and spinel inorganic materials: A review. Renewable and Sustainable Energy Reviews. 2022;**157**:112047

[12] Wategaonkar SB, Parale VG, Pawar RP, Mali SS, Hong CK, Powar RR, et al. Structural, morphological, and optical studies of hydrothermally synthesized Nb-added TiO_2 for DSSC application. Ceramics International. 2021;**47**:25580-25592

[13] Geleta TA, Imae T. Nanocomposite photoanodes consisting of p-NiO/n-ZnO heterojunction and carbon quantum dot additive for dye-sensitized solar cells. ACS Applied Nano Materials. 2022;**4**:236-249

[14] Alabdan HI, Alwabsi A, Alsagrey AM, AlKhulaif FH, Abdelgawad ME, AlQahtani NJ, et al. Usage of CuO in dye-sensitized solar cells and CuO-based dyesensitized solar cells: A review. IEEE Journal of Photovoltaics. 2022;**12**:1445-1452

[15] Chalkias DA, Giannopoulos DI, Kollia E, Petala A, Kostopoulos V, Papanicolaou GC. Preparation of polyvinylpyrrolidone-based polymer electrolytes and their application by in-situ gelation in dye-sensitized solar cells. Electrochimica Acta. 2018;**271**:632-640

[16] Salih EY, Bashir MBA, Rajpar AH, Badruddin IA, Bahmanrokh G. Rapid fabrication of NiO/porous Si film for ultra-violate photodetector: The effect of laser energy. Microelectronic Engineering. 2022;**258**:111758

[17] Li C, Xin C, Xu L, Zhong Y, Wu W. Components control for high-voltage quasi-solid state dye-sensitized solar cells based on two-phase polymer gel electrolyte. Solar Energy. 2019;**181**:130-136

[18] Abisharani JM, Dineshkumar R, Devikala S, Arthanareeswari M, Ganesan S. Influence of 2, 4-diamino-6-phenyl-1-3-5-triazine on bio synthesized TiO_2 dye-sensitized solar cell fabricated using poly (ethylene glycol) polymer electrolyte. Materials Research Express. 2020;**7**(2):025507

[19] Kotatha D, Hirata M, Ogino M, Uchida S, Ishikawa M, Furuike T, et al. Preparation and characterization of electrospun gelatin nanofibers for use as nonaqueous electrolyte in electric double-layer capacitor. Journal of Nanotechnology. 2019

[20] Salih EY, Ramizy A, Aldaghri O, Mohd Sabri MF, Madkhali N, Alinad T, et al. In-depth optical analysis of Zn (Al) O mixed metal oxide film-based Zn/Al-layered double hydroxide for TCO application. Crystals. 2022;**12**:79

[21] Bashir MBA, Salih EY, Sabri MFM, Rajpar AH, Badruddin IA, Hussein MZ, et al. In-depth thermal, microstructural and photoluminescence analysis of mesoporous ZnO/ZnAl2O4-MMO: The effect of molar ratio. ECS Journal of Solid State Science and Technology. 2021;**10**:106006

[22] Salih EY, Sabri MFM, Tan ST, Sulaiman K, Hussein MZ, Said SM, et al. Preparation and characterization of $ZnO/ZnAl_2O_4$-mixed metal oxides for dye-sensitized photodetector using Zn/Al-layered double hydroxide as precursor. Journal of Nanoparticle Research. 2019;**21**:1-12

[23] Balamurugan S, Ganesan S. Novel cobalt redox materials admitted in natrosol polymer with a thiophene based additive as a gel polymer electrolyte to tune up the efficiency of dye sensitized solar cells. Electrochimica Acta. 2020;**329**:135169

[24] Qin H, Owyeung RE, Sonkusale SR, Panzer MJ. Highly stretchable and nonvolatile gelatin-supported deep eutectic solvent gel electrolyte-based ionic skins for strain and pressure sensing. Journal of Materials Chemistry C. 2019;**7**(3):601-608

[25] Liu S, Liu J, Wang T, Wang C, Ge Z, Liu J, et al. Preparation and photovoltaic properties of dye-sensitized solar cells based on zinc titanium mixed metal oxides. Colloids and Surfaces A: Physicochemical and Engineering Aspects. 2019;**568**:59-65

[26] Foruzin LJ, Rezvani Z, Nejati K. TiO2@NiAl-Layered double oxide nanocomposite: An excellent

photoanode for a dye sensitized solar cell. Solar Energy. 2019;**186**:106-112

[27] Ge Z, Zhu Y, Wang C, Xia L, Guo L, Wu Y, et al. Investigation of the photoanode based on graphene/zinc aluminum mixed metal oxide for dye-sensitized solar cell. Journal of Sol-Gel Science and Technology. 2020;**95**:432-438

[28] Fasolini A, Sangiorgi N, Brandi ET, Sangiorgi A, Mariani F, Scavetta E, et al. Increased efficiency and stability of dye-sensitized solar cells (DSSC) photoanode by intercalation of Eosin Y into Zn/Al layered double hydroxide. Applied Clay Science. 2021;**212**:106219

[29] Tan CY, Omar FS, Saidi NM, Farhana NK, Ramesh S, Ramesh K. Optimization of poly (vinyl alcohol-co-ethylene)-based gel polymer electrolyte containing nickel phosphate nanoparticles for dye-sensitized solar cell application. Solar Energy. 2019;**178**:231-240

[30] Pullanjiot N, Swaminathan S. Enhanced electrochemical properties of metal oxide interspersed polymer gel electrolyte for QSDSSC application. Solar Energy. 2019;**186**:37-45

[31] Yang Y, Zhang Z, Gao J, Pan D, Yuan B, Guo X, et al. Metal-organic materials as efficient additives in polymer electrolytes for quasi-solid-state dye-sensitized solar cells. Journal of Alloys and Compounds. 2017;**726**:1286-1294

[32] Chang WC, Sie SY, Yu WC, Lin LY, Yu YJ. Preparation of nano-composite gel electrolytes with metal oxide additives for dye-sensitized solar cells. Electrochimica Acta. 2016;**212**:333-342

[33] Mazloum-Ardakani M, Arazi R, Haghshenas M, Tamaddon F, Alizadeh M. Synthesis of 2-amino-4-(4-(methylamino) phenyl)-6-phenylnicotinonitrile as a new additive for the passivation of the TiO_2 surface and retarding recombination in dye-sensitized solar cells. Electrochimica Acta. 2018;**266**:452-459

[34] Nath NCD, Jun Y, Lee JJ. Guanidine nitrate (GuNO3) as an efficient additive in the electrolyte of dye-sensitized solar cells. Electrochimica Acta. 2016;**201**:151-157

[35] Riaz R, Ali M, Maiyalagan T, Arbab AA, Anjum AS, Lee S, et al. Activated charcoal and reduced graphene sheets composite structure for highly electro-catalytically active counter electrode material and water treatment. International Journal of Hydrogen Energy. 2020;**45**:7751-7763

[36] Bagheri O, Dehghani H, Afrooz M. Pyridine derivatives; new efficient additives in bromide/tribromide electrolyte for dye sensitized solar cells. RSC Advances. 2015;**5**(105):86191-86198

[37] Abraham N, Unni C, Philip D. Studies on bandgap tuning of visible light active heterojunction CuO/ZnO nanocomposites for DSSC application. Journal of Materials Science: Materials in Electronics. 2019;**29**:21002-21013

[38] Rajakumar P, Satheeshkumar C, Ravivarma M, Ganesan S, Maruthamuthu P. Enhanced performance of dye-sensitized solar cell using triazole based phenothiazine dendrimers as additives. Journal of Materials Chemistry A. 2013;**1**(44):13941-13948

[39] Godlewski MM, Kaszewski J, Kielbik P, Olszewski J, Lipinski W, Slonska-Zielonka A, et al. New generation of oxide-based nanoparticles for the applications in early cancer detection and diagnostics. Nanotechnology Reviews. 2020;**9**:274-302

[40] Salih EY, Ramizy A, Aldaghri O, Sabri MFM, Madkhali N, Alinad T, et al. Rapid synthesis of hexagonal-shaped Zn (Al) O-MMO nanorods for dye-sensitized solar cell using Zn/Al-LDH as precursor. Nanomaterials. 2022;**12**:1477

[41] Musa A, Ahmad MB, Hussein MZ, Saiman MI, Sani HA. Effect of gelatin-stabilized copper nanoparticles on catalytic reduction of methylene blue. Nanoscale Research Letters. 2016;**11**(1):1-13

[42] Zhuang C, Tao F, Cui Y. Anti-degradation gelatin films crosslinked by active ester based on cellulose. RSC Advances. 2015;**5**(64):52183-52193

[43] Karthika P, Ganesan S, Arthanareeswari M. Low-cost synthesized organic compounds in solvent free quasi-solid state polyethyleneimine, polyethylene glycol based polymer electrolyte for dye-sensitized solar cells with high photovoltaic conversion efficiencies. Solar Energy. 2018;**160**:225-250

[44] Karthika P, Ganesan S, Kamalakannan S, Prakash M. Design and synthesis of the D–π–A-structured coadsorbents with the phenanthraquinone core and its application in dye-sensitized solar cells. The Journal of Physical Chemistry C. 2020;**124**(18):9886-9899

[45] Pavithra N, Velayutham D, Sorrentino A, Anandan S. Poly (ethylene oxide) polymer matrix coupled with urea as gel electrolyte for dye sensitized solar cell applications. Synthetic Metals. 2017;**226**:62-70

[46] Selvaraj B, Shanmugam G, Kamaraj S, Gunasekeran A, Sambandam A. Effect of 1-substituted 2-(pyridin-2-yl)-1 H-Benzo [d] imidazole ligand-coordinated copper and cobalt complex redox electrolytes on performance of Ru (II) dye-based dye-sensitized solar cells. Inorganic Chemistry. 2021;**60**(3):1937-1947

[47] Nagaraj P, Sasidharan A, David V, Sambandam A. Effect of cross-linking on the performances of starch-based biopolymer as gel electrolyte for dye-sensitized solar cell applications. Polymers. 2017;**9**(12):667

[48] Khan A, Alamry KA. Recent advances of emerging green chitosan- based biomaterials with potential biomedical applications: A review. Carbohydrate Research. 2021;**506**(5):108-368

[49] Marques ADS, Da Silva VAS, Ribeiro ES. Dye-sensitized solar cells: Components screening for glass substrate, counter-electrode, photoanode and electrolyte. Materials Research. 2020;**23**(5):e20200168